THE NEW GROVE

Second Viennese School

SCHOENBERG

WEBERN   BERG

Oliver Neighbour
Paul Griffiths
George Perle

M

PAPERMAC

MACMILLAN LONDON

First published in
The New Grove Dictionary of Music and Musicians,
edited by Stanley Sadie, 1980

First published in paperback with additions 1983 by
PAPERMAC
a division of Macmillan Publishers Limited
London and Basingstoke

Associated companies in Auckland, Dallas,
Delhi, Dublin, Hong Kong, Johannesburg,
Lagos, Manzini, Melbourne, Nairobi,
New York, Singapore, Tokyo, Washington
and Zaria

ISBN 0 333 35384 6

Printed in Hong Kong

Contents

Alban Berg   *George Perle*

List of Illustrations

We are grateful to the following for permission to reproduce illustrative
material: Universal Edition (Alfred A. Kalmus Ltd.), Vienna (cover, figs.2, 5, 7,
10, 13, 14); Belmont Music Publishers, Los Angeles (cover, figs.1, 4, 6); Verlag
Fritz Molden, Vienna (fig.3); Pierpont Morgan Library, New York (fig.9);
Historisches Museum der Stadt Wien (fig.11); Mosco Carner, London (fig.12).

General Abbreviations

A	alto, contralto [voice]	orch	orchestra
acc.	accompaniment	orchd	orchestrated
B	bass [voice]	org	organ
b	bass [instrument]	perc	percussion
b	born	perf.	performance, performed by
Bar	baritone	pic	piccolo
bn	bassoon	qt	quartet
cel	celesta	qnt	quintet
cl	clarinet	red.	reduction, reduced for
cond.	conductor	repr.	reprinted
db	double bass	rev.	revision
ens	ensemble	S	soprano [voice]
fl	flute	sax	saxophone
gui	guitar	str	string(s)
hn	horn	sym.	symphony, symphonic
inc.	incomplete	T	tenor [voice]
inst	instrument	t	tenor [instrument]
Jg.	Jahrgang [year of publication/volume]	tpt	trumpet
		trans.	translation
		trbn	trombone
mand	mandolin	unpubd	unpublished
Mez	mezzo-soprano	v, vv	voice(s)
movt	movement	v., vv.	verse(s)
n.d.	no date of publication	va	viola
		vc	cello
ob	oboe	vn	violin

Symbols for the library sources of works, printed in *italic*, correspond to those used in *RISM*, Sec. A.

Bibliographical Abbreviations

AMw	*Archiv für Musikwissenschaft*
BMw	*Beiträge zur Musikwissenschaft*
CMc	*Current Musicology*
DAM	*Dansk aarbog for musikforskning*
DJbM	*Deutsches Jahrbuch der Musikwissenschaft*
DTÖ	*Denkmäler der Tonkunst in Österreich*
HV	*Hudebni věda*
IMSCR	*International Musicological Society Congress Report*
JAMS	*Journal of the American Musicological Society*
JMT	*Journal of Music Theory*
Mf	*Die Musikforschung*
ML	*Music and Letters*
MM	*Modern Music*
MMA	*Miscellanea musicologica*
MMC	*Miscellanea musicological*
MO	*Musical Opinion*
MQ	*The Musical Quarterly*
MR	*The Musical Review*
MT	*The Musical Times*
NRMI	*Nuova rivista musicale italiana*
NZM	*Neue Zeitschrift für Musik*
ÖMz	*Österreichische Musikzeitschrift*
PNM	*Perspectives of New Music*
PRMA	*Proceedings of the Royal Musical Association*
RaM	*La rassegna musicale*
ReM	*La revue musicale*
RIM	*Rivista italiana di musicologia*
RMI	*Rivista musicale italiana*
SM	*Studia musicologica Academiae scientiarum hungaricae*
SMA	*Studies in Music [Australia]*
SMz	*Schweizerische Musikzeitung/Revue musicale suisse*
ZMw	*Zeitschrift für Musikwissenschaft*

Preface

This volume is one of a series of short biographies derived from *The New Grove Dictionary of Music and Musicians* (London, 1980). In its original form, the text was written in the mid-1970s, and finalized at the end of that decade. For this reprint, the text has been re-read and modified by the original authors and corrections and changes have been made. In particular, an effort has been made to bring the bibliographies up to date and to incorporate the findings of recent research.

The fact that the texts of the books in the series originated as dictionary articles inevitably gives them a character somewhat different from that of books conceived as such. They are designed, first of all, to accommodate a very great deal of information in a manner that makes reference quick and easy. Their first concern is with fact rather than opinion, and this leads to a larger than usual proportion of the texts being devoted to biography than to critical discussion. The nature of a reference work gives it a particular obligation to convey received knowledge and to treat of composers' lives and works in an encyclopedic fashion, with proper acknowledgment of sources and due care to reflect different standpoints, rather than to embody imaginative or speculative writing about a composer's character or his music. It is hoped that the comprehensive work-lists and extended bibliographies, indicative of the origins of the books in a reference work, will be valuable to the reader who is eager for full and accurate reference information and who may not have ready access to *The New Grove Dictionary* or who may prefer to have it in this more compact form.

S.S.

SCHOENBERG

Oliver Neighbour

CHAPTER ONE

Life

I 1874–1914

Arnold Franz Walter Schönberg (or Schoenberg, to use the spelling which he adopted when he emigrated to America) was born in Vienna on 13 September 1874. His father Samuel (1838–90) was born in Szécsény, his mother (née Nachod, 1848–1921) in Prague. They came to Vienna from Pressburg (Bratislava). Schoenberg accordingly inherited Hungarian nationality, which was converted to Czech on the formation of the state of Czechoslovakia in 1918. He became an American citizen in 1941. The family was Jewish, and the three children, Arnold, Ottilie and Heinrich, were brought up in the orthodox faith. Neither parent was particularly musical; Schoenberg remembered his uncle Fritz Nachod, who wrote poetry and taught him French, as the main cultural influence of his childhood. But his sister and brother showed musical talent, and the latter, like their cousin Hans Nachod, became a professional singer. Schoenberg's musical education began when he was eight with violin lessons, and he very soon began composing by the light of nature, imitating the violin duets by such composers as Pleyel and Viotti that he was given to learn, and arranging anything that came his way – operatic melodies or military band music – for the same combination. Somewhat later, having met a schoolfellow who played the viola, he was able to spread his wings to the point of writing trios for two violins and viola.

The family was not well off. In the year after the death of his father, who had kept a shoe shop, Schoenberg was obliged to leave school and take employment as a clerk in a small private bank, where he remained for about five years. Meanwhile he pursued music, literature and philosophy in the evenings, his interest fired by two friends of his own age, David Josef Bach and Oskar Adler. According to his own account Bach taught him the courage to keep his artistic ideals high. Adler was in effect his first music teacher. He was a good violinist, and Schoenberg taught himself the cello, at first using a large viola adapted with zither strings, and then a proper cello which he began by playing with violin fingering. Together they formed an amateur ensemble which permitted Schoenberg to explore the Classical chamber music repertory from the inside and to compose quartets. Adler helped him to educate his ear through playing, and taught him some elementary harmony. For the musical forms he turned to articles in a popular encyclopedia.

Schoenberg and his friends heard very little music except what they could play themselves. Concerts were beyond their means, though they would sometimes stand outside café enclosures to eavesdrop on the band. While he was still working in the bank Schoenberg joined an amateur orchestra, really no more than a handful of string players, conducted by Alexander von Zemlinsky, and the two soon became firm friends. Zemlinsky, the elder by two years, had attended the Vienna Conservatory, where he had distinguished himself. His compositions had attracted Brahms's notice. He was therefore in a position to help Schoenberg with the formal instruction that he had so far missed. Although ·

Schoenberg received encouragement from Josef Labor, to whom he submitted a movement from a string quartet in C in about 1894, and from Richard Heuberger, Zemlinsky was the only regular teacher he ever had. The importance of Zemlinsky's influence is hard to assess. In later life Schoenberg ascribed to him most of his knowledge of the problems and techniques of composing, whereas Zemlinsky merely said that they had shown each other their works. It is difficult to believe that Schoenberg ever needed to be prompted twice about a general principle of composition, but he certainly respected Zemlinsky's advice, and the pattern of their early relationship persisted. At a time when misunderstanding had taught him to hold himself aloof he still treated Zemlinsky as an equal both as man and musician.

In the autumn of 1897 Schoenberg wrote a string quartet in D major, making various changes in the course of composition in response to Zemlinsky's criticisms. When it was done both felt that it marked a new stage in his work, and Zemlinsky, who was on the committee of the Wiener Tonkünstlerverein, proposed it for performance. It was accepted, played at a concert for members only the following March, and well enough received to be repeated in the next season. It was many years before a new work of Schoenberg's was to meet with comparable success. The Verein turned down his string sextet *Verklärte Nacht* in 1899, and there were protests when songs from opp.1–3 were sung in public in December 1900. From that time on, in his own words, the scandal never stopped. In these early works he had already taken the first steps in the development of chromaticism that was to lead him to abandon triadic

1. Fröhliches Quintet, c1895: the cellist is Schoenberg and the violinist with the moustache Fritz Kreisler

harmony and tonality itself by 1908, and each stage in his progress aroused fresh hostility. For the moment, however, little was heard of him. He kept the wolf from the door by conducting small choral societies and orchestrating operettas, and managed between March 1900 and April 1901 to compose the vast *Gurrelieder*.

In October 1901 Schoenberg married Zemlinsky's sister Mathilde (1877–1923). There were two children of the marriage: Gertrud (1902–47), who married Schoenberg's pupil Felix Greissle in 1921 and emigrated to the USA in 1938, and Georg (1906–74). In December the young couple moved to Berlin, where Schoenberg had got a job on the musical side of Überbrettl, a kind of cabaret that formed part of Ernst von Wolzogen's Buntes Theater. The idea behind Überbrettl was to use the popular mode to serious ends. Various well-known men of letters, such as Wedekind, Morgenstern and Dehmel, were interested in it. In the summer Schoenberg had tried his hand at setting verses of the Überbrettl type, and at least one song, *Nachtwandler*, was subsequently performed in Berlin, though only once. Schoenberg's employment there lasted only until the following summer, after which he was obliged to interrupt the orchestration of the *Gurrelieder* in order to score operettas. He was saved from further drudgery of this kind by Richard Strauss, to whom he had shown parts of the *Gurrelieder* and his new symphonic poem *Pelleas und Melisande*. Strauss was impressed, and used his influence to obtain for him the Liszt Stipendium and a post as composition teacher at the Stern Conservatory. So he stayed on in Berlin for another year and returned to Vienna in July 1903 with the completed score of *Pelleas*.

That autumn various musical classes were organized in rooms made available at a girls' school founded by Dr Eugenie Schwarzwald. Schoenberg taught harmony and counterpoint there for a single season, and Zemlinsky, in whose house he was living at the time, form and orchestration. When Schoenberg gave up his class some of its members continued to study composition and theory with him privately, among them a number of students of music history under Mahler's friend Guido Adler at the University of Vienna. In the autumn of 1904 this nucleus was joined by two new recruits, Webern (an Adler pupil) and Berg, who were to fulfil their promise as composers through acceptance and individual reinterpretation of the successive steps in their master's development, and bring him the support of their lifelong personal and artistic loyalty.

If private teaching was scarcely lucrative for Schoenberg – he taught Berg free for the first year because his family was not in a position to pay fees – composition was still less so. The Viennese public was conservative in its tastes and reluctant to support new work in any of the arts. Special societies attempted to remedy this situation. To one of them, the Ansorge Verein, Schoenberg owed various early performances, starting with some of his songs early in 1904. At this time he and Zemlinsky were already planning a society of their own, which they launched successfully under the title Vereinigung Schaffender Tonkünstler. For their honorary president they managed to secure Mahler, whose brother-in-law Arnold Rosé had invited him to rehearsals of *Verklärte Nacht* the previous year when Rosé was preparing the quartet that he led for a performance of it. Mahler was deeply impressed and became a staunch supporter of Schoenberg, even though he did not always see eye to eye with him over artistic matters. The new society survived only for the season 1904–5 but succeeded in putting on sizable works by Mahler, Strauss, Zemlinsky and others, and in January the first performance of *Pelleas und Melisande*, conducted by the composer. The orchestra was ill at ease and the reception cool.

The pattern of Schoenberg's life for the next few years was now set. A heavy teaching programme did not save him and his family from material hardship; as late as 1910 he was obliged to borrow from Mahler to pay the rent, and the following year Berg launched an appeal on his behalf, though without his knowledge. The style of his music, which he composed largely in the slacker summer months, became increasingly dissonant; each new work raised a storm. The Rosé quartet gave the first performances of the First Quartet and First Chamber Symphony early in 1907. Mahler stood up for both works in public, and although he privately confessed that he could not fully understand Schoenberg's development he never lost faith in him. His removal from Vienna that spring deprived Schoenberg of a valuable ally, though in the four years that remained to him his concern for Schoenberg's well-being and interest in his work never faltered. Uproar predictably greeted Rosé's first performance of the Second Quartet in December 1908, and when the first freely dissonant works, *Das Buch der hängenden Gärten* and the op.11 piano pieces, were presented in January 1910 they met with almost universal incomprehension.

During these years of crisis for his musical style Schoenberg also turned to painting, and in October

1910 mounted a one-man exhibition. The following January he received a letter from the expressionist painter Kandinsky, whose sympathy for his work extended beyond his painting to his music and ideas. This initiated a lasting friendship. Schoenberg exhibited with the group Der Blaue Reiter founded by Kandinsky, and contributed an essay and a facsimile of *Herzgewächse* to the first and only number of the periodical that bore its name. He showed pictures elsewhere, but, although he continued to paint and draw occasionally in later years, visual means of expression quickly lost the importance that they had briefly held for him.

For some years Schoenberg had kept up a fairly steady output of music, culminating in the extraordinary works of 1909: the op.11 piano pieces, the Five Orchestral Pieces op.16 and *Erwartung*. But now the pace slackened. His spare time in the years 1910–11 was largely devoted to writing the *Harmonielehre* and completing the long-delayed orchestration of the *Gurrelieder*. In 1910 he offered his services to the Royal Academy of Music and Dramatic Art as an external lecturer in theory and composition. His application was successful, but his hopes that this might lead to a professorship were thwarted. A question was asked in parliament, and he was subjected to virulent attacks on racial grounds. By the end of the academic year his circumstances had so far deteriorated that he decided to try his luck once again in Berlin, and moved there with his family in the autumn of 1911.

His arrival was greeted with some extremely unpleasant comment in the press, and his winter lectures at the Stern Conservatory were poorly attended. Nevertheless his fortunes at last began to improve a little. His name at least was now internationally familiar, audiences were beginning to find his earlier music more accessible, and his later work was arousing curiosity. *Pierrot lunaire*, composed in the summer of 1912, was given with considerable success under the composer's direction in October, and then went on tour to 11 German and Austrian cities. Sir Henry Wood had given the first performance of the Orchestral Pieces in London the previous month, and that of the *Gurrelieder* took place in Vienna the following February under Schreker. This was an overwhelming success, but the composer, smarting under years of very different treatment from the Viennese public, refused to acknowledge its applause. Five weeks later it took its revenge by bringing a concert of music by Schoenberg and his associates to a halt. Meanwhile Schoenberg, relieved of immediate financial worries by the generosity of a rich patron, determined to make a secondary career as a conductor. He lacked experience, but Zemlinsky arranged for him to conduct, early in 1912, a concert including *Pelleas und Melisande*. This set him on the road. By the outbreak of war he had conducted *Pelleas*, the *Gurrelieder* and the Orchestral Pieces in a number of European cities.

II World War I and after

The outbreak of World War I put an end to these developments. Concerts, especially those involving new music, were less in demand. Many of Schoenberg's pupils were called up, and his teaching ceased entirely. In May 1915 he was himself medically examined in Vienna for the reserve, but to his surprise he was rejected on account of goitre. In September he moved his family back to Vienna, having accepted after some hesitation the offer of a rent-free house from his patron Frau Lieser. Then, after a second medical examination had reversed the decision of the earlier one, he finally joined up in December as a one-year volunteer. Schoenberg's health had, however, never been strong; under the strain of a course of training at Bruck an der Leitha he began to suffer from asthma, to which he was subject all his life, and other ailments. Friends tried to secure his release, which came through quite un-expectedly in October 1916. In the last four years he had written very little music, apart from finishing *Die glückliche Hand* in 1913 and composing the Four

2. Arnold Schoenberg

Orchestral Songs op.22 at intervals between that year and 1916. But he had been constantly preoccupied with plans for a large-scale religious work. After his return to civilian life he finally decided to embody his ideas in an oratorio. By May 1917 the text of *Die Jakobsleiter* was ready.

In June he began to compose the music. The time could scarcely have been less favourable. Food and the coal necessary to cook it were becoming desperately short in Vienna; money, at least in the Schoenberg household, was shorter still. Yet in the space of three months Schoenberg set the whole of the first part of the oratorio, though without fully working out the orchestration. During the same period he made known plans for a seminar in composition which would avoid any set course of instruction unrelated to the individual needs of the pupil, and for which each pupil would pay only what he could afford. September brought further difficulties. Schoenberg found himself obliged to leave his house. Potential landlords showed themselves suspicious of his prospects, and for many weeks the family endured the acute discomfort of cheap boarding-houses. On September 17 he was called up again. This time he was given C grading, and, although a transfer away from Vienna remained a possibility until his final discharge in December, his duties were much lighter than before and he was often at home. Consequently he was able to go forward with his seminar at the Schwarzwald school. It prospered, and after his move to Mödling the following April he continued to hold classes there till 1920. But to the oratorio the short spell of military service proved fatal. Despite constant efforts to pick up the thread, he had managed by 1922 to compose only about half of the interlude intended to link the two halves of the work, after which he added nothing more.

A direct outcome of the seminar was the foundation of a Society for Private Musical Performances, the object of which was to give properly rehearsed performances of modern works to a genuinely interested membership. For one class of seat members paid only according to their means. The press was excluded. Details of programmes were not available in advance, and many works were repeated as a point of policy. Orchestral works were given in arrangements for piano or chamber ensemble. In the three years between February 1919 and the end of 1921, when inflation put an end to the society's activities, 353 performances of 154 works were given in 117 concerts. A number of Schoenberg's pupils and ex-pupils helped with the organization of this vast enterprise, but he rehearsed and directed a considerable proportion of the performances himself. Meanwhile peace brought a renewal of international interest in his music. Conducting engagements took him abroad. In Amsterdam he was made president of the International Mahler League, and he returned there for the winter of 1920–21 to take part in a festival of his own works and give a series of lectures on music theory. This was the time of the formulation of serialism. The first three serial works, the op.23 piano pieces, the Serenade and the Piano Suite, were written between 1920 and 1923. The Wind Quintet was completed the next year, which saw the first performances not only of the Serenade and Quintet, but of *Erwartung* (in Prague) and *Die glückliche Hand* (in Vienna).

In October 1923 Mathilde Schoenberg died. Although the marriage had run into difficulties in earlier years Schoenberg's letters written at the time of her death leave no doubt of the depth of his attachment to her. A month later he completed his *Requiem*, a meditation on death the first section of which had been drafted somewhat earlier; he never set it to music. His widowerhood did not, however, last long: at the end of the following August, about a fortnight before his 50th birthday, he married Gertrud Kolisch, the sister of his pupil Rudolf Kolisch. (Kolisch was a violinist and the leader of a string quartet which became the leading exponent of Schoenberg's chamber music in the 1920s and 1930s.) There were three children of this marriage: Nuria (*b* Barcelona, 1932), who married the Italian composer Luigi Nono, Rudolf Ronald (*b* 1937) and Lawrence Adam (*b* 1941).

In 1925 Schoenberg was invited to take charge of the master class in composition at the Prussian Academy of Arts in Berlin, in succession to Busoni, who had died the year before. He accepted, signed the contract in September, and after some delay because of an appendix operation moved in January 1926 from Vienna to Berlin for the third and last time. Some of his pupils, notably Gerhard and Zillig, moved with him, and Eisler, though no longer his pupil, did so independently at about the same time; Skalkottas was to join the class a little later. For the next seven years Schoenberg enjoyed better conditions of work than at any time in his life. He had a say in general questions of policy and administration in the academy, and absolute responsibility for his own courses. Moreover he was required to teach for an average of only six months in the year, and could choose his own times. His creative output increased correspondingly. The Suite op.29, largely written in Vienna, was followed by the Variations for orchestra, the play *Der biblische Weg*, the Third Quartet, *Von heute auf morgen*, the *Begleitmusik zu einer Lichtspielszene*, *Moses und Aron*, the Cello Concerto after Monn, and various smaller pieces. His earlier works continued to gain ground with audiences, and his more recent ones were at least assured of a hearing, if not of approval: the Variations, for instance, had a very mixed reception when Furtwängler introduced them in 1928.

Given that Schoenberg could never hope to make a living from composition, his job at the academy was well adapted to his needs. Perhaps in the long run he would not have stood the climate of Berlin, for in the winter of 1930–31 his asthma grew much worse, and he made so little progress in the summer that he was strongly advised not to risk the next winter in the north. So in October the Schoenbergs went to Barcelona to stay near Gerhard and his wife; various circumstances kept them there until May. However, it was not Schoenberg's health but politics that robbed him of any sense of security in Berlin. Antisemitism had contributed considerably to the hostility towards him in Vienna even before the war. In the early 1920s, when he experienced the grossly insulting behaviour towards Jews that Hitler's agitation was helping to make commonplace, he already foresaw violence as the probable outcome. By 1933 the realization of his fears had begun. It was no surprise when the government's intention to remove Jewish elements from the academy was announced at a meeting of the senate on 1 March, at which Schoenberg was present. He left abruptly, and treated the announcement as his dismissal. This took effect officially from the end of October, in breach of his contract, which should have protected him for another 23 months.

The Schoenbergs left Berlin in May and spent the summer in France. The only work composed at this time was the String Quartet Concerto after Handel. One of Schoenberg's first acts was to return to the Jewish faith, which he had rejected in favour of Lutheranism in 1898. His Christian beliefs had not lasted, but by his own account he was at no time unreligious, let alone anti-religious. By the war years religion had become his sole support. At first he did not attempt to reconcile his beliefs with those of any recognized faith, but with the increase of antisemitism after the war he realized that the faith in which he had been brought up must eventually claim him, and he began to work his way towards his own not entirely orthodox version of it. The ceremony in Paris merely made his reconversion official.

III **America**

Schoenberg's search for employment finally ended with his acceptance of a teaching post until the following May at the Malkin Conservatory in Boston. The family arrived in the USA at the end of October. The work proved to be on a more elementary level than he had realized. Some of the classes were held in New York, which meant a tiring weekly journey there. As soon as the weather became bad in December his health deteriorated; he fell seriously ill in January and again in March. The summer put him right, but he dared not stay another winter on the east coast and moved to Los Angeles in the autumn of 1934 for the sake of the climate – a decision that probably added several years to his life. He first settled in Hollywood, where he completed the Suite for string orchestra by the end of the year. Private pupils soon began to come to him, and in the academic year 1935–6 he gave lectures at the University of Southern California. In 1936 he accepted a professorship in the University of California at Los Angeles, and moved to a house in Brentwood Park where he lived for the rest of his life. That year saw the composition of the Fourth Quartet and the completion of the Violin Concerto, apparently begun the previous spring or summer.

Though more fortunately placed in his country of exile than many of his fellow refugees, Schoenberg enjoyed little peace of mind. He found much in his alien surroundings hard to accept; few of his pupils were well enough grounded to benefit at all fully from his knowledge and experience; there was no audience for such music as he might write; above all there was the appalling news from Europe and the growing threat to relatives and friends there. His constant efforts on behalf of individual victims of persecution could not ease the sense of helplessness of one who was accustomed to take remedies into his own hands. For once he admitted to depression. In due course, however, he made some kind of truce with his situation. The war disposed in its own way of certain issues. His domestic happiness was a source of strength, and his young American children gave him a certain stake in the country. In the four years after 1936 his only original works had been *Kol nidre*, intended for synagogue use, and the completion of the Second Chamber Symphony, partly composed between 1906 and 1916; but in 1941 he composed the Organ Variations in response to a commission, and three more works had followed by 1943. He also set about recasting material from various unfinished theoretical works in the form of a series of more strictly practical textbooks suitable for his American pupils. Nevertheless, in 1944 he was still thinking of emigrating.

This year was a turning-point in two respects. In February his health began to deteriorate sharply. Diabetes was diagnosed, he suffered from giddiness and fainting, and his asthma grew worse, as did the optical disturbances that had troubled him for some time. On reaching his 70th birthday in September he had to give up his professorship. As he had taught in the university for only eight years his pension was very small. Consequently he was obliged to continue giving private lessons, and in 1946 held a course of lectures at the University of Chicago. In August that year he had a heart attack which caused his heart to stop beating; he was resuscitated only by an injection directly into the heart. This experience is in some sense reflected in the String Trio which he composed shortly after his recovery. Although he was well enough in the summer of 1948 to give classes at Santa Barbara, for most of his remaining five years he led the withdrawn existence of an invalid. But he had the satisfaction of seeing the emergence of the state of Israel (he was elected honorary president of the Israel Academy of Music in 1951), and also the upsurge of interest in his music that marked the postwar years. At this time he revised a small selection from his vast accumulation of largely unpublished essays and articles, and published it under the title *Style and Idea*. The few short compositions that he managed to complete were nearly all religious in inspiration.

During the last year of his life he worked on a series of meditations which he originally called *Modern Psalms*, and later *Psalmen, Gebete und Gespräche mit und über Gott*; his last composition was an incomplete setting of the first of these. He died in Los Angeles on 13 July 1951.

IV Personality and beliefs

The scanty recollections of those who knew Schoenberg in early years stress his enthusiasm and resilience. Such qualities are only to be expected in a young man just finding scope for uncommon gifts, yet one circumstance behind Schoenberg's growing confidence during the decade before *Verklärte Nacht* claims attention for its fundamental influence on his later outlook and thinking: the fact that he was in all essentials self-taught.

Fortune had endowed him not only with prodigious musical aptitude but with the intellectual energy and force of personality to ensure that it triumphed over his very considerable social and educational disadvantages. Naturally he took what steps he could to make up for his lack of formal musical training, but neither his haphazard reading, nor other odd crumbs of instruction (he is known, for instance, to have heard Bruckner lecture at the academy), nor even Zemlinsky's constant help, could alter his feeling that he never profited from what he was taught unless he had already discovered it for himself; tuition could at best only awaken him to his own knowledge. The process of independent discovery shaped his habits of mind and his spiritual life. His approach to composition, whether in the context of a single work or of his wider development, remained exploratory; he saw life as synonymous with change and religion as a quest.

His early experience is most closely reflected in, and so partly deducible from, his teaching methods. He refused to teach the codified knowledge that he had never learnt, mistrusting mere knowledge as the enemy of understanding. From the earliest stages his pupils were required to create, to derive their simplest exercise from an expressive intention and to remain true to the implications of the initial idea. Their teacher let no inconsequence pass, just as at a deeper level he would detect any transgression against the promptings of their musicality. For many of Schoenberg's pupils, particularly in the earlier years, the kind of moral obligation that he taught them to feel towards the demands of their art found an echo in their whole attitude to life, and they grouped themselves round him like a band of disciples. Their master benefited from the relationship too, for the origin of his lifelong interest in teaching lay in the need constantly to re-enact his own exploration of the resources of music. Just as many composers, himself among them, might exercise their contrapuntal skill in canonic problems, Schoenberg, who habitually thought in terms of processes rather than systems, practised the ability to reach outwards from a given starting-point by helping each pupil to work out his own salvation in accordance with his own personality and musical disposition.

It might be supposed that this approach to teaching would have led to great stylistic freedom, especially in view of his condemnation in the *Harmonielehre* of all academic rules as meaningless abstractions from the practice of a past era. However, he taught strictly within the confines of tonality, and made the principles of traditional grammar live again by demonstrating their functional value for his pupils' work as for that of the great Austrian and German composers, whom he constantly called to witness. His points of departure for technical instruction – Sechter in the *Harmonielehre* and *Structural Functions of Harmony*, Fux in *Preliminary Exercises in Counterpoint*, Classical forms in *Models for Beginners in Composition* and *Fundamentals of Musical Composition* – were relatively unimportant: everything depended on reinterpretation, on exploration through trial and error. His primary aim was to teach logical thinking, and that was best done in a context where theory, which must necessarily lag behind practice, could aid elucidation. Here again his teaching reflects his own position as a composer, which he was at

pains to clarify in the *Harmonielehre*. He was convinced
that the recent developments in his style, although
reached intuitively, were a logical outcome of tradition,
and that, while taking no account of rules, they observed
fundamental laws which would eventually prove defin-
able. Meanwhile the pupil who felt drawn to similar
modes of expression must find his own intuitive path
with the aid of self-reliance learnt in better-charted territory, and the listener would need faith.

In the crucial years preceding the *Harmonielehre*
Schoenberg's music rarely met with faith or even the
modicum of goodwill without which no artistic perception is possible. On the contrary, it was opposed with
almost unbelievable persistence and venom. Perhaps no
music before or since has encountered such a reception;
to the end of his life its author, though internationally
famous, had to accept very widespread incomprehension. The price he paid for artistic integrity was propor-
tionately high. It should be remembered that the sense of
outrage that even such a work as *Pelleas und Melisande*
aroused at first in the majority of listeners arose not
only from unthinking conservatism but from the more
positive instinct that its premonitions of a radical dis-
ruption in the agreed basis of musical language carried a
threat to precision of meaning. Schoenberg, who shared
his audience's background and many of its assumptions,
understood its fears and so experienced its attack with
something like the force of an inner doubt, requiring all
the more courage to parry. He felt himself impelled
towards the break with tonality almost despite himself,
and accomplished it only after considerable hesitation.
Since its systematic justification in theory eluded him he
looked for some other authority to protect his intuition.

21

He found it eventually in religion.

In the year after Mahler's death in 1911 Schoenberg wrote about him in terms that indicate clearly his pre-occupations at that time. He attacked with great bitterness those whose ceaseless denigration of Mahler almost led him to lose faith in his own work, and apostrophized him as saint and martyr. He saw all great music as expressing the longing of the soul for God, and genius as representing man's more spiritual future, so that the uncomprehending present must inevitably persecute the good and promote the bad. His quotation of Mahler's remark that the Eighth Symphony was composed at great speed, almost as though from dictation, is

4. Self portrait of Schoenberg, c1910

especially significant, for he too composed very quickly, often with the feeling that however much effort he put into his work something more was given that he could not account for, just as his stylistic development seemed to have been taken out of his hands. It was not only Mahler and his great predecessors whom he had come to see as divinely inspired: his admission that the role of the 'chosen one' in *Die Jakobsleiter* was based on his own experience removes any doubt that he placed himself in their company. (However, Mahler's music never influenced his own at all deeply, and his sympathy for it sometimes wavered – to his discomfort, because he linked entitlement to respect with the ability to accord it.)

Schoenberg's need to understand his artistic role can scarcely have been the only factor in the spiritual crisis that led to his rediscovery of religious faith: it is merely the one to which his work and writings give access. Similarly the ideas embodied in the prose drama *Der biblische Weg* and in *Moses und Aron* cannot fully document the return to Judaism as a result of which religion became his support in racial as in artistic persecution. The decision to make this return official proved a difficult one because it seemed to set the seal on his divorce from the Western tradition which had nurtured him and to which he had contributed so powerfully. In reaction he even spoke at the time of giving up composition and devoting himself to the Jewish national cause. That did not happen, but his personal and racial idealism remained closely intertwined to the end of his life, as a letter written within three months of his death to the Israel Academy of Music shows:

23

Those who issue from such an institution must be truly priests of art, approaching art in the same spirit of consecration as the priest approaches God's altar. For just as God chose Israel to be the people whose task it is to maintain the pure, true, Mosaic monotheism despite all persecution, despite all affliction, so too it is the task of Israeli musicians to set the world an example of the old kind that can make our souls function again as they must if mankind is to evolve any higher.

The idea of the artist as priest or prophet is often deprecated as inflated, complacent, arrogant or presumptuous. But no reader of *Die Jakobsleiter* and *Moses und Aron* will imagine that Schoenberg looked for cheap self-justification or easy solutions to spiritual or artistic problems. The path that had been pointed out to him was unmarked, to be followed blindfold and often with anguish, in the knowledge that it would be lost the moment faith faltered. Moreover the need to protect the supremacy of faith came into conflict with the urge to rationalize and justify: faith must fear conscious constraints yet needed the support of discipline, which must accordingly in some sense cross the divide between the rational and the intuitive. This ultimately irresoluble tension ran all through Schoenberg's thinking and showed itself in many guises. It lies, for instance, at the heart of serialism, where every note is brought within the law, but in such a way that intuition retains its freedom. And an analogous dichotomy provides the subject of *Moses und Aron*, which concerns the simultaneous duty and impossibility of giving expression to inexpressible truths.

Unhappily Schoenberg's struggle to realize his ideals dominated not only his spiritual but his social life, where the humility belonging to the former too often deserted him. He could not ignore misunderstanding, but fought back. As he said himself in a letter of 1924:

Unfortunately the better sort of people become enemies faster than friends because everything is so serious and important to them that they are perpetually in a defensive position. They are driven to this by the great, indeed ruthless honesty with which they treat themselves and which makes them adopt the same attitude to other people as well. It is very wrong, really, for we human beings are far too much in need of tolerance for any thoroughgoing honesty to be helpful to us. If only we could manage to be wise enough to put people on probation instead of condemning them, if we could only give proven friends such extended credit! – I am speaking of my own defects, knowing very well why I have often been more lonely than could well be pleasant.

Even here he seems to miss the implication of his habitual insistence on his place among 'better' people: to expect respect is to discourage it even in those who recognize that it is due. He did not make life easy for his adherents, regarding interest in modern music beyond that of his own circle as betrayal. No doubt it was true that the contemporary listener or performer prepared to devote himself wholeheartedly to Schoenberg's music would have found it almost as difficult as the composer himself to sympathize with other modes of thought, but he must sometimes have driven away genuine well-wishers along with the opportunists. His enjoyment of his months in Barcelona in 1931–2 arose partly from relief at escaping from the pedestal that he had built for himself in Berlin, and being accepted as an equal by people who knew little about him.

Readers of Schoenberg's posthumously published correspondence, however, discover not only his less accommodating side but much that only the more fortunate of his contemporaries could know: his absolute honesty in all his dealings, his generosity of mind wherever he sensed integrity, his delicacy of feeling where he saw the need to temper his customary directness, his energy in expressing sympathy through practical help, his capacity for gratitude, his loyalty. His critical and aesthetic writings, turning as they invariably do on matters that concern him deeply, reveal his personality no less vividly, displaying the same rather lofty yet compelling idealism, the same irascible pride, the same flashes of humour and warmth, the same justice within the framework of strongly held convictions. His thinking here is at all times a creator's, never that of the historian concerned to give everything its place. He is content to speak as an individual, with a more selfconscious view of his relation to tradition than his predecessors enjoyed, but still with the confidence of one who knows where he stands. Integrity of personality enables limitations in his historical sympathies, and even inconsistencies in the logic on which he naively though not unjustifiably prided himself, to fall into place beside his unique insights into the music that he valued and the musical crisis in which he found himself involved. The special perceptions that distinguish his writing arise directly out of his experience in composition, and so, it would seem, does his manner of presentation, at once direct and cogent yet unexpected and elliptical. And that is hardly surprising, since it is in music that his mind and spirit found their fullest expression.

CHAPTER TWO

Works

1 Early tonal works

Schoenberg's music may be divided into four periods, the second and third of which were inaugurated by crises in compositional technique that had important consequences not only for the composer's own work but for music in general. The music of the first period is tonal, or at least employs a tonality as a central point of reference. In 1908 Schoenberg abandoned tonality; he was the first composer to do so. The music of the ensuing second period is often called 'atonal'. Schoenberg considered this term nonsensical, preferring 'pantonal'. Since either term properly embraces his serial music as well, the period will be referred to here as 'expressionist'. From his work of this time he gradually evolved the principle of serialism, which he first used consistently in 1920; the serial music written between that date and 1936 constitutes the third period. The fourth, less well defined phase may be said to emerge during the 1930s. It is marked by greater stylistic diversity, including occasional returns to tonal composition.

Of the music that Schoenberg is known to have composed in large quantities from childhood to his early 20s not very much survives, and some of that is fragmentary. Unfinished pieces remained with the composer, whereas completed ones were played with friends and usually lost. Songs have fared best, though some of the larger unfinished works contain complete movements. Although Schoenberg had not yet acquired the habit of dating his manuscripts, it should eventually prove possible to trace his early development, at least in outline. At the time of writing, however, only three works antedating his op.1 have been published: two sets of piano pieces, one each for solo and duet dated 1894 and 1896 respectively, and the D major String Quartet of 1897.

The piano pieces scarcely hint at their composer's future stature, but they already display characteristic preoccupations. The solo pieces are fairly ambitious ternary structures, no doubt inspired by Brahms's sets of the previous two years. They show a good grasp of the possibilities offered at the lead-back and coda, but clumsy execution not helped by uncertain feeling for piano textures. Attempts at more original effects – the links between the coda of each piece and the beginning of its successor, and the descant melody in diminution in the third piece – sound distinctly forced. Heuberger, to whom Schoenberg showed some songs at about this time, advised him to write some short pieces in the style of Schubert. The six little duet pieces were the result. Schoenberg clearly took the point that he must learn complete control by testing his every step. He subjugated himself to the same discipline that half a century later he was still advocating in *Models for Beginners in Composition* and *Fundamentals of Musical Composition*. Each melody progresses by drawing on its own motivic resources, which also permeate the accompaniment, and the consequences of every harmony are carefully weighed. The pieces (except no.5) are arranged in ascending order of formal development. The first consists simply of two repeated eight-bar strains. In each subsequent piece there is a little more expansion after the double bar, culminating in tiny contrasting episodes in nos.4 and 6. Only in these two pieces is the slightest deviation from four-bar phraseology admitted. Throughout his life, and especially after 1920, Schoenberg's music drew strength from his acute sensitivity to phrase structure, shifts of emphasis within a regular rhythmic framework and the tensions arising from asymmetry. In the duets he set about sharpening a faculty that the solo pieces show to have been innate.

There is no reason to suppose that these sets of pieces would appear specially important in Schoenberg's output of the mid-1890s if more of it were known. The D major String Quartet, however, marks a huge stride forward, and not only on the available evidence: the composer himself recognized it as a turning-point and remembered it with affection. Brahms is still the dominant influence. The work owes its Classical four-movement layout to his mediation, its structural cogency and clarity derive from him, and so to a large extent does the style, though certain themes speak with a strong Czech accent. Yet there is a freedom of movement, a deftly guided fluency, that does not belong to the older master's closely considered manner, and it is here that Schoenberg's musical personality asserts itself most strikingly. His sheer zest in the making of music is one of his most persistent characteristics: it accounts for the feeling of resilience that accompanies his exploration of even the darkest regions of experience and tempers his findings. If the D major Quartet, delightful though it is, does not seem fully typical of him it is due less to the eclectic idiom than to the absence of another constant factor in his music: the sense of urgency in communicating a particular conception.

This quality, however, begins to make itself felt in the pair of lengthy songs which Schoenberg wrote almost certainly in the following year and eventually selected as his op.1. The effort to match the magniloquent sentiments of the verses called forth better things from the young composer than they deserved. True, the naivety that prompted the choice of text comes through, rather endearingly, in the setting. But although the Wagnerian influence that was to loom so large in the next few years is already perceptible, there is no close model for the firm sonata-influenced forms, the wealth of independent contrapuntal development in the accompaniments or the distinctive breadth and warmth of the asymmetrical melodic lines.

Schoenberg found inspiration for several compositions of 1899 in poems by Dehmel: the songs *Warnung*, *Erwartung*, *Erhebung* and presumably *Schenk mir deinen goldenen Kamm*, and the string sextet *Verklärte Nacht*. (The charming *Waldsonne* of about this time is not a Dehmel song and stands apart.) The desire to give expression to the feelings aroused in him by Dehmel's work considerably influenced the development of his style, as he later confessed to the poet. The songs are shorter and reach out to sharper, less generalized experience than those of op.1, though the main preoccupation is still love. Their concentration of means and mood shows one kind of advance, the expansive textures of *Verklärte Nacht*, in which Wagnerian and Brahmsian modes of thought meet in harmonious accord, a contrasting one. In the Dehmel poem that served as the basis for this symphonic poem a woman confesses to her lover that she is already pregnant by another man, and he replies that through their love the child will be born his own. A knowledge of this unlikely tale is of secondary importance to the listener because the lack of action enables the work to be understood as a single-movement abstract composition. No composer understood better than Schoenberg that music serves its subject best when claiming for itself the greatest possible autonomy.

In March 1900 Schoenberg began setting Jens Peter Jacobsen's *Gurrelieder* as a song cycle for voice and piano, for entry in a competition. In accordance with the ballad-like tone of the verse he built the vocal lines from relatively simple rhythmic elements, a style shared by the songs *Hochzeitslied* and *Freihold* (the first probably and the second certainly dating from the same year), and perhaps suggested by some of Zemlinsky's early songs. However, Schoenberg soon saw wider possibilities in the text. Having fallen under Wagner's spell he felt the need for subjects that transcended common experience, his first thought being to wring something more from such well-worn themes as love, death and transfiguration. The way lay through mastery and reinterpretation of Wagnerian style, and the *Gurrelieder* offered a far more expansive arena for this important confrontation than *Verklärte Nacht* had done. He therefore decided to connect the songs he had already composed (those in the first two parts of the finished work) with symphonic interludes and set the whole poem as a vast cantata employing several soloists and a huge chorus and orchestra.

The work depicts the love of King Waldemar and Tove under the Tristanesque imminence of death,

Waldemar's blasphemous defiance of God after Tove's death, the nightly ride at the head of a ghostly retinue to which the king's restless spirit is subsequently condemned, and its dismissal by the summer wind at the approach of day. Schoenberg encompassed all this in a series of tableaux of extraordinary magnificence. But the poem deals with dramatic events in an undramatic form and so required some kind of interpretative emphasis to bring the great musical design clearly into focus. The opportunity was there, for at some level Schoenberg's choice of the poem must surely have been influenced by Waldemar's rebellion against God and the renewal brought about as the summer wind sweeps away the aftermath of human passion – both themes that border on his religious concerns of a few years later. Yet neither emerges with unifying force, whether because he was unable to commit himself fully to the text or through inexperience in dramatic matters. As late as 1913 he could still write to Zemlinsky that he did not consider himself a dramatic composer in the ordinary sense. In the *Gurrelieder* he tended to fall back on direct reminiscence of Wagner's later operas, especially *Götterdämmerung*, to evoke atmosphere or characterize events. It is significant that after considering an opera on Maeterlinck's *Pelléas et Mélisande* for his next work (he knew nothing of Debussy's opera), he rejected the idea in favour of a symphonic poem on the same subject.

Schoenberg later said that it was Maeterlinck's ability to lend timelessness to perennial human problems that had attracted him to the play. Certainly it was precisely the moments least involved with the action that inspired him to step furthest outside his own chronology towards his stylistic future, for instance in the music associated with Mélisande's first mysterious appearance heard at the outset and again before her death. But such music as Golaud's, and that of the main love scene, is less advanced; it is capable of traditional extension, notably through Wagnerian sequence, and therefore well adapted to carry the narrative. The contrapuntal virtuosity surpasses even that of the *Gurrelieder*, constantly changing the expressive colour of the thematic material in a manner that is entirely individual while paying tribute to Wagner – rather than to Strauss, whose influence appears sporadically on a more superficial level. Yet for all its riches the work contains a structural conflict. The Mélisande and Pelléas themes lose something of their essence as they are drawn into the larger contrapuntal development, a process that may fit the symbolism of the work but also suggests that the composer had not yet mastered the potentialities of his more striking inventions.

Schoenberg now returned to songwriting. The songs of the next three years fall into three groups. Those of the first group, dating from 1903 and the earlier months of 1904, explore various subjects. *Wie Georg von Frundsberg* and *Das Wappenschild*, a fiery showpiece with orchestra, follow the lead of *Freihold* as songs of defiance. They must surely contain the composer's reaction to hostility; perhaps the gloomy *Verlassen* does so too in a different way. *Die Aufgeregten* reflects ironically on human passion, though love remains the theme of some of the most beautiful of these songs. *Geübtes Herz*, *Traumleben* and the orchestral *Natur* cultivate the intense lyrical style first heard in *Schenk mir deinen goldenen Kamm*. *Ghasel* continues this line, but with a change of emphasis in the accompaniment, which

33

involves the voice part in imitation and adopts its even flow. The three Petrarch sonnets from op.8, composed in the later part of 1904 when the D minor String Quartet was already under way, form a distinct group set a little apart from Schoenberg's other songs. Their contrapuntal style derives directly from *Ghasel*, but takes a far more complex form made possible by the orchestral setting.

In the third group, dating from 1905, the vocal lines regain their independence, relying on motifs rather than imitation to relate them to their accompaniments. Except for the slightly earlier orchestral *Sehnsucht* all these songs, which are based on a curious assortment of serious and trivial verses, were composed about the time of the completion of the D minor Quartet, and already show the characteristics of Schoenberg's tonal thinking in its last stages. His early liking for chromatic approaches to diatonic notes, strikingly manifested as early as *Erwartung* (1899), had led to ever-increasing chromatic substitution, especially in the melodic field. This in turn required clarification by correspondingly elaborate harmonization, employing so wide a range of primary and altered degrees within the tonality that modulation lost its force. So his music, which had at no time inclined to constant modulation, became increasingly monotonal. This tendency appears in all the songs, but in two contrasting forms: in *Der Wanderer*, *Am Wegrand* (later quoted in the opera *Erwartung*) and *Mädchenlied*, as in *Verlassen* of 1903, the tonal centre is strongly, sometimes almost obsessively stressed, whereas in *Sehnsucht*, *Alles* and *Lockung* it is scarcely touched on.

The D minor Quartet, Schoenberg's first wholly characteristic and assured large-scale masterpiece, consists, like *Pelleas und Melisande*, of a single vast movement, but naturally without illustrative interludes. A scherzo, slow movement and rondo are interspersed at various points between the first part of the development and the coda of what would normally have been the first movement, and absorbed into it by the use of common material. The general idea for such a form originates in Liszt, whose novel formal concepts Schoenberg admired while finding his attempts to put them into practice schematic and unfelt. But the quartet arose more directly from Schoenberg's fundamental preference for abstract composition reasserting itself and acting upon his recent cultivation of the Straussian symphonic poem. The twin formative influences of Wagner and Brahms once again find an even balance, as they had in *Verklärte Nacht*, but now completely and finally assimilated. Perhaps the most striking single quality of this work is its extraordinary melodic breadth. As the melodies move away from their initial, firmly tonal contexts, develop, and combine contrapuntally, they form what Schoenberg called vagrant harmonies; the music, though not very dissonant, loses tonal definition. Thus the structure cannot be understood entirely in tonal terms. Its powerful sense of direction is maintained through the composer's exceptional capacity to shape his material in relation to its formal purpose, a capacity that after his abandonment of tonality was to prove strong enough to carry a far heavier structural burden. Late in life he remarked that he had never been content to introduce an idea for structural reasons alone: it must always make a positive contribution to the substance of the work. The D minor Quartet already displays the typical Schoenbergian richness fostered by this habit of mind.

The First Chamber Symphony, completed in July 1906, adopts the quartet's single-movement layout, but in a more concise form; though in no way a slighter work it is barely half as long. Schoenberg aimed here at concentration rather than expansiveness and, as he was so often to do in solving the problems posed by a particular conception, opened up possibilities for the future remote from his immediate artistic concern. In the first place he increased his instrumental forces from four to 15 in order to accommodate the simultaneous presentation of a greater concentration of ideas. Viewed from another angle, however, the increase appears as a reduction: it established the soloistic orchestral writing already found here and there in the *Gurrelieder* and *Pelleas*, and opened the way for the small, strongly differentiated instrumental ensembles appropriate to Schoenberg's later style – and that of many younger composers. But the urge towards concentration affected deeper levels in his musical thought. The two opening themes are based respectively on superimposed perfect 4ths and the whole-tone scale, both of which readily form chordal structures. The distinction between the melodic and harmonic dimensions thus becomes blurred, a process closely bound up with the loss of tonality in Schoenberg's music. However, for the moment the E major frame held.

Although the imminence of change may seem obvious to the listener with hindsight, it was not so to the composer. On completing this exuberant work he felt that he had now arrived at a settled style. The music of the next year or so reflects this conviction. In neither the eight-part chorus *Friede auf Erden*, which he later described as an illusion written when he still thought harmony among men conceivable, nor in the two Ballads op.12, does the threat to tonality grow appreciably. Schoenberg always regretted that he had not had time to follow up all the implications of the style of this period, and 30 years later returned to the task. For the present, however, some inner crisis urged him towards new realms of expression and hastened the inevitable revolution. A change of mood had made itself felt earlier. The songs of 1905, for instance, provide an uneasy, questioning interlude between the confident First Quartet and First Chamber Symphony, and the Second Chamber Symphony, begun immediately after the first, opens in a new spirit of sombre resignation. Despite repeated attempts he was unable to finish this work at the time, perhaps because he could not reconcile the more carefree spirit in which the second movement opens with his changing preoccupations. At all events it was a very intimate, elusive piece, the contemporary first movement of the Second Quartet, that spoke for him now and demanded to be followed up.

II **Expressionist works**

The new quartet did not, however, occupy his whole attention; it was not finished until the later months of 1908. At the same time he wrote songs and took up painting seriously. This unexpected development apparently arose out of his personal contact with the Viennese painters Oskar Kokoschka and Richard Gerstl. By far the greater part of his work in this sphere belongs to the years 1908–10, when his music underwent its first great crisis. The pictures are mostly portraits or strange, imaginary heads – 'visions' as he called some of them. They are amateurish in execution yet sufficiently skilful to convey the intensity of his imagination, and it seems likely that their importance to him lay in this very opposition. This was a time when artists and writers who were later to be called expressionists sought to obey the promptings of the spirit ever more directly, in some sense bypassing the machinery of artistic tradition in order to reach deeper levels of experience. The relation to tradition remained, of course, the crucial factor: as a painter Schoenberg's amateur status severely limited the scope and quality of his achievement, but allowed him to feel that his hand was guided without his conscious intervention, whereas in music he had to pay for the benefits of mastery by reckoning with its censorship. So for a time his method of painting represented the ideal towards which his real work of composition aspired.

In the winter of 1907–8 Schoenberg interrupted work on the scherzo of the Second Quartet to compose the two songs op.14. They are highly imitative pieces, the second reminiscent of *Ghasel* in texture. As in some of the songs of 1905 tonic harmony scarcely appears until the close and now exerts still less gravitational force. Certain dissonances, notably perfect and altered 4th chords, resolve so tardily and so variously as to weaken expectation of their doing so at all. This process reached its logical conclusion shortly afterwards in songs from *Das Buch der hängenden Gärten*, at least five of which (nos.4, 5, 3, 8, 7) are known to date from March and April 1908. Here dissonance is finally emancipated, that is, it no longer seeks the justification of resolution. Consequently structural harmony disappears, along with its need for measured periods and consistent textures, and so does tonality itself as a central point of reference. By way of compensation motivic work and the tendency to equate the horizontal and vertical dimensions – in fact the essential elements later codified in the serial method – assume greater responsibility. The poems by George that led Schoenberg to explore the untried expressive possibilities of free dissonance describe in rather indirect language the growth of a passion in an exotic setting and the subsequent parting. Neither poet nor composer wishes to arouse sympathy or evoke ecstasy. The songs are predominantly slow and quiet, the lack of tonal or rhythmic propulsion placing them outside time. Each one captures with peculiar vividness the shifts of feeling at a particular moment, but distanced, as though enshrined in the limbo of past experience. There is nothing of Waldemar and Tove here: the summer wind will assuredly soon sweep all before it. What will be left? Schoenberg gave his answer in the Second Quartet, one of the most personal of all his works.

This quartet consists of four thematically related movements which successively reflect the transformation of his style, but do not further it. The third movement is later but less advanced than the op.14 songs, and the finale, though tonal only in parts, stands in the same relation to the earlier songs of *Das Buch der hängenden Gärten*. The reason for this lies not only in the technical consideration that the later movements could not overstep certain limits set by the enigmatic first movement, which is in F♯ minor: the composer needed to step back in order to see the crisis that had overtaken him clearly. For the crisis was the subject of the work. The trio of the scherzo incorporates the popular melody *O du lieber Augustin*, the words of which end with the tag 'Alles ist hin', as a private reference to his wife's liaison with Gerstl. The two later movements contain settings for soprano voice of George poems, the first a prayer for divine solace after earthly struggles, the second a vision of the spirit's journey to ethereal realms. Although Schoenberg's choice of subject for his next vocal works was to be directed towards human insights, he evidently recognized already that his ultimate aim was religious.

Early in 1909 Schoenberg composed the first two piano pieces of op.11, before completing *Das Buch der hängenden Gärten*. The Five Orchestral Pieces op.16 and the third piece of op.11 followed in the summer. The strange note of resignation that had sounded through the song cycle is still heard in op.11 nos.1 and 2 and op.16 no.2, but the unfamiliar territory of the new style now takes in the explosive turmoil of op.16 nos.1 and 4 and op.11 no.3, and the unique calm of op.16 no.3. Formal expansion does not accompany the extension of expressive range: as Schoenberg later observed, brevity and intensity of expression are interdependent in these pieces. The disintegration of functional harmony appeared at the time to have destroyed the conditions for large-scale form. But other features with roots in traditional practice, in particular fixed points of reference of various kinds (some of them reminiscent of tonality) and thematic or motivic development, survived to assume not only greater responsibility but new guises. These made possible swifter transformations and more abrupt contrasts than music had hitherto known. Moreover dissonance's new independence permitted, at least in an orchestral context, unprecedented simultaneous contrasts. It is not only novelty of expression in itself but the power to bring seemingly irreconcilable elements into relation that gives the music its visionary quality, far beyond that of the painted 'visions'

For a time Schoenberg believed that by following the dictates of expression he would be able to renounce motivic features as well as tonality. The last two pieces in opp.11 and 16 to be written, the final piece in each set, show the direction of his thinking. The orchestral piece centres on a continuously evolving melodic line with no clear expository stage; the piano piece relies for coherence as much on dynamics and texture as on pattern. From this point two possibilities suggested themselves. One was to devise ideas that were complete in themselves and required no development. This held no lasting attraction for a composer of Schoenberg's imaginative fecundity. He composed two tiny pieces for chamber ensemble and part of a third early in 1910, and the next year six equally minute piano pieces which he published as op.19; thereafter he left this line of thought to Webern. For Schoenberg the way forward lay in the construction of large forms on the basis of a text. This allowed him scope to build on the experience of opp.11 and 16. Immediately after the instrumental pieces he composed in the astonishingly short time of 17 days the half-hour monodrama *Erwartung*.

The single character in this piece is an unnamed woman. Full of fear and apprehension, she is wandering through a forest at night in search of her lover. The only dramatic event, her discovery of his murdered body, occurs at a fairly early stage; the rest of her monologue passes from recollection of their love, through jealousy to a sense of reconciliation born of exhaustion. As the composer remarked, the whole drama may be understood as a nightmare, but the point is immaterial because the reality explored is purely psychological. There is no realistic time scale: past and present co-exist and merge in the woman's mind as terror, desire, jealousy and tenderness cut across one another in confused association. Traditional tonal order could scarcely have met the demands of such a subject: Schoenberg's extraordinary score depends to a considerable extent upon a rationality beyond conscious control. True, various unifying factors are observable, such as fixed pitch elements that turn upon a vestigial D minor (his favourite key throughout his life, whether in tonal, freely pantonal or serial composition) and a number of motivic figures that recur time and again, especially at the beginning of phrases. But since these are short, widely scattered and quickly submerged in the stream of continuous development their contribution to coherence at surface level is small; the music can scarcely be called athematic, but it goes further in that direction than any other work of Schoenberg. The monologue falls into several lengthy paragraphs which provide the clearest structural feature, but even here divisions are blurred and larger changes of mood disrupted by innumerable contradictory emotions. Beyond a certain point nothing can impinge upon the dreamlike continuum of musical images.

The next year, 1910, Schoenberg wrote the text of *Die glückliche Hand,* and began the music soon after, though he did not finish it for three years. It is a companion-piece to *Erwartung*, in effect another monodrama, centring on an unnamed man. Though shorter it requires more elaborate staging, including an intricate play of coloured lighting synchronized with the action. The subsidiary roles – a woman, a gentleman and some workers – are mimed, since they are merely projections of the man's psyche, but the chorus of 12 soloists, whose commentary opens and closes the drama, reveals through its pity of him that it represents an independent, presumably divine order of existence. At the beginning the chorus asks why he constantly betrays his capacity for the supermundane in a vain quest for earthly happiness. The main action symbolizes this situation. The man loves a woman who deserts him for a rival, but seems to return to him. In the mistaken belief that he has won her he finds strength to withstand his enemies and inspiration for artistic creation. His resulting work is symbolized by a trinket; it excites envy, but he recognizes it as meretricious. The woman plays him false and the cycle is complete. Although the style of the music is close to that of *Erwartung* Schoenberg reintroduces features that he had temporarily set aside, to meet the more varied action and the wider implications of the text. Clear formal divisions reassert themselves: recapitulatory reminiscence plays an important part in the later stages of the action and there are correspondences between the flanking choral scenes, where exact imitation reappears. There is also a new element, barely hinted at in the works of 1909: the use of parody to characterize such situations as the metal working and the woman's fickleness.

Parody assumes a very important role in *Pierrot lunaire*. This work, composed in 1912, before the completion of *Die glückliche Hand*, consists of 21 poems set

5. *Autograph MS of the opening of 'Rote Messe' from 'Pierrot lunaire', composed 1912*

for speaker and chamber ensemble. Schoenberg had employed melodrama before in the summer wind narrative of the *Gurrelieder*, and the choruses of *Die glückliche Hand* are also partially spoken. His highly stylized use of the speaking voice, for which he notated relative pitches as well as exact rhythms, proved an ideal vehicle for the Pierrot settings, which were conceived in what he described as a light, ironic–satirical tone. The rather modish verses, by turns grotesque, macabre or consciously sentimental, provide the occasion for presenting, with the detachment that the protagonist in *Die glückliche Hand* failed to achieve, human activity as a shadow play in which menace and absurdity are on a level. The focus shifts at random, as in a dream, between the lunatic activities of the clown, impersonal scenes, the poet in the first person and the self-absorbed artist, who is not spared. Within his new style Schoenberg parodies the characteristics of a great range of genre pieces, very often retaining the ghost of their formal layout as well. In music the lines dividing ironic from direct reference are often hard to detect. The peculiar fascination of *Pierrot lunaire* lies in this ambiguity. The nightmare imagery of some of the poems might scarcely be admissible without ironic distancing, yet the music often strikes with authentic horror. Mockery constantly shades into good humour, exaggerated pathos into the genuinely touching. A decade later Schoenberg was to rediscover his sympathy for the world that he was now determined to leave behind him. For the moment, however, he was set on other things.

After *Pierrot* Schoenberg contemplated writing an oratorio based on the vision of Swedenborg's Heaven at the end of Balzac's novel *Séraphita*. This idea was superseded during 1914 by plans for a vast, partly choral symphony of a religious nature, incorporating texts from Dehmel, Tagore and the Old Testament. Early in 1915 he wrote words for a new final section consisting of two movements entitled *Totentanz der Prinzipien* and *Die Jakobsleiter*, but although he made extensive sketches nothing came to fruition until he decided to make his own statement of faith by turning *Die Jakobsleiter* into an independent oratorio. He began to revise the text in 1916 and composed the first half the next year. At the beginning of the allegory, which owes

45

a good deal to Balzac's *Séraphita*, a host of people approaching death come before the archangel Gabriel, who admonishes and advises them. Six representatives of various philosophical standpoints then come forward to recount their earthly experiences and aspirations, and receive his comments. There is no doubt something of Schoenberg in all of them, and in Gabriel too, but he avowedly identified himself with the 'chosen one', whose spiritual understanding sets him apart and whose word seems doomed to misunderstanding. A central symphonic interlude symbolizing the transition from this world to the hereafter leads to the uncomposed second part in which souls are prepared for reincarnation as the next step in their long spiritual pilgrimage towards ultimate perfection. The chosen one is reluctant to face the world again, once more to stand alone and find himself involuntarily compelled, though receiving no support, to speak and do what he would never have dared to think or take responsibility for. But he is told to remember all that he has in common with the rest of humanity and to accept his prophetic role. At the close Gabriel calls on every soul to seek unity with God through prayer.

The faith and the view of his mission to which Schoenberg gave expression in *Die Jakobsleiter* were to influence the whole course of his later development as a composer. The short score of the first part, however, is more easily seen as a potential culmination to the music composed since 1908 than as a foretaste of that of the 1920s. The closing section of *Die glückliche Hand* provided a model for the big, partly sung, partly spoken choruses. The long paragraphs sung or spoken by the soloists required a more sustained style of writing than that, for instance, of *Erwartung*, where the varying intensity of dissonance breaks continuity of pace and texture. For this Schoenberg was able to turn to the Four Orchestral Songs op.22 (1913–16) and their forerunner *Herzgewächse* (1911), where he had already devised more even textures by maintaining a rather high level of dissonance in six or more parts, with very little octave doubling and a tendency towards symmetrically built chords. Except for *Seraphita* (op.22 no.1) all these songs anticipate the religious preoccupation of *Die Jakobsleiter*. The very high soprano voice that symbolizes prayer in *Herzgewächse* reappears as the soul that ascends heavenward just before the central interlude. In January 1915 Schoenberg wrote to Zemlinsky that his new symphony would be 'worked' ('ein gearbeitetes Werk') in contrast to his many 'purely impressionistic' recent works. He carried this resolve over into the oratorio. His brief exploration of the dream world of free association had permanently enriched his musical language and vision, but he now needed to regain greater formal elaboration and density of meaning. Although *Die Jakobsleiter*, like the monodramas before it, relies primarily on the text for its structure, it employs recurrent themes and melodies, often in contrapuntal combination. Many of these are related through permutations of a hexachord heard at the outset.

At one point in the unfinished central interlude Schoenberg directed that groups of instruments placed at a distance should enter in 'floating' ('schwebend') rhythm not exactly synchronized with that of the main orchestra. The suspension of rhythmic propulsion symbolizes the dissolution of earthly ties on the threshold of the hereafter. How far Schoenberg would have been able to pass beyond this extraordinary conception into the Swedenborgian Heaven of his text had he not been interrupted, it is impossible to say, though the history of *Moses und Aron* suggests that he would not have reached the end. But the 12-note serial method that increasingly occupied him from 1920 provided a continuation of another sort. The omnipresent series sought to establish as principles the equation of the horizontal and vertical aspects of music, and the unity of all ideas in a composition with each other and with their context. Schoenberg expressly compared the unity of musical space to Swedenborg's concept of Heaven where 'there is no absolute down, no right or left, forward or backward'. In a different sense from the symphonic interlude the music must 'float'. The dodecaphonic aspect counteracted the pull of tonal gravity; the only quasi-tonal music in *Die Jakobsleiter* belongs to 'one of the called', who is roundly rebuked for preferring beauty to truth. In June 1922, shortly before he gave up trying to continue the oratorio, but when his foot was already firmly on the serial path, Schoenberg started a new sketchbook by inscribing the cover with the words 'Mit Gott'.

III Serial and tonal works 1920–36

Since serialism is a compositional method and does not dictate style, he might have been expected to find in it the means, if not of finishing *Die Jakobsleiter*, at least of continuing on a path suggested by that work. Instead he evolved a form of neo-classicism. This may not have been his original intention. His first known attempt at consistent serial composition is a dodecaphonic orchestral passacaglia for which he made sketches in March 1920. Its economic textures look forward to the

Serenade and Wind Quintet (and he remembered the introduction six years later when he began the Variations for Orchestra), but the structural conception seems to show affinities with the music for 'one who strives' in *Die Jakobsleiter*. The Piano Pieces op.23 nos.1, 2, and 4, written or begun in July 1920, are descendants of the pre-war instrumental pieces, and he began the Piano Suite op.25, exactly a year later, with the only two movements (the Prelude and Intermezzo) that are not dance movements, thinking of the work simply as a second set of pieces. However, the Variations and Dance Scene from the Serenade op.24 had been begun in the later months of 1920, the March followed in September 1921, and by the time all three works were finished in the early part of 1923 movements based on Classical forms predominated. Although every piece in opp.23 and 24 involves serial procedures, only one in each work uses a 12-note series. Both of these postdate the earliest movements of the Suite, which, like the abandoned passacaglia and nearly everything that Schoenberg was to compose in the next ten years, is dodecaphonic throughout.

The reason for Schoenberg's return to Classical forms must be sought in his need to find new scope for his inherently developmental cast of thought. Paradoxically, developing variation had brought about, above all in the later works of 1909, a reduction in the conditions for its own exercise. Where every motif is transformed before it can gather associations for the listener there can be no intensification of meaning through development; where no pattern establishes itself only extreme contrasts cheat expectation, and then not for long. If Schoenberg's art of development was to develop further it needed a basis in relative stability, especially in the rhythmic sphere. For him technical needs were inseparable from philosophical ones. It seems likely that he saw his music at this time as initiating a new incarnation analogous to that required of the 'chosen one' in the second part of *Die Jakobsleiter*. In the second turn of the spiral of his musical existence his task was evidently to reinterpret, in accordance with the 'higher and better order' to which he aspired, not his own previous experience, but the course of musical history as he knew and understood it best. His real interest began with Bach. He later declared his teachers to have been in the first place Bach and Mozart, and in the second Beethoven, Wagner and Brahms. Although the last two had appeared as the dominant influences in his tonal music, at least on the surface, the earlier ones now came to the fore. Despite the reluctance of the 'chosen one', like Moses after him, to return to the world and prophesy, Schoenberg was able to write to Hauer in December 1923 that after a 15-year search he had discovered a method of composition that allowed him to compose with a freedom and fantasy such as he had only known in his youth. The next 13 years were remarkably fruitful.

Most of the movements in the Serenade and the Piano Suite draw on late Baroque dance characteristics much as *Pierrot lunaire* had borrowed from the subjects that it parodied. But although the detail of the Serenade often recalls *Pierrot*, as does its humour, six of its seven movements are built on an altogether larger scale, even without the lengthy repeats that Schoenberg adopted from his models. The repeats, here and in the Piano Suite, are the first of any size and almost the last in the whole of his published work. They set him the special problem of canalizing his transforming imagination sufficiently within a given mood and character for a repetition to make sense. The exercise was no doubt an essential step towards establishing strongly differentiated developing characters in the great instrumental and operatic structures of the coming years. But that was incidental: Schoenberg said that he never knew what lay ahead, and his zigzag course towards the crises of 1908 and 1920 bears him out. There is nothing merely preparatory about the early serial masterpieces: his concern was, as ever, with the unique work in hand.

Thus in the marvellous series of instrumental works composed between 1920 and 1936 individuality is not of the limited kind associated with stepping-stones in a stylistic or technical evolution. In each one vigorous expansion within the terms of a particular premise builds a self-sufficient statement of very wide range, yet entirely singular. The next two works, the Wind Quintet and the Suite op.29 for seven instruments, illustrate the point very clearly. Schoenberg turned here to the thematic contrast required by Classical forms and to the traditional four-movement pattern. The first movement of the Quintet follows standard sonata layout, and the finale is a rondo. The first movement of the Suite lacks a regular development section, but despite the dance character of the second and fourth movements consistent symphonic treatment allies it with the Quintet rather than the Serenade. Yet the two works differ radically. The persistent contrapuntal texture of the Quintet looks back to the First String Quartet and First Chamber Symphony (and the emphasis on whole-tone and quartal sonorities is reminiscent of the latter work); the Suite is rooted in a harmonic idea which pervades texture and melody throughout. The divergence affects the music at every level.

6. *Arnold Schoenberg*

In the Variations for orchestra (1926–8) and the
Third String Quartet (1927), which are also modelled
on Classical forms, Schoenberg avoided these contrapuntal and harmonic extremes for the most part, and
finally established the main stylistic characteristics of
his serial music; these were to remain fairly constant to
the end of his life. The transformations of the series as

such cannot, of course, be followed consistently by the ear, and he strongly deprecated any attempt to do so. Although for him the series functioned in the manner of a motif, his themes consist primarily of rhythmic patterns which may carry any serial derivation. The thematic rhythms themselves are not fixed: he showed remarkable skill in varying them without endangering their identity. The interplay of melodic and rhythmic motif is responsible to a very large extent for the extraordinary richness of the music, bringing about in the course of a work the gradual accumulation of a mass of affinities between disparate elements. It also affects the bar-to-bar texture in an important way. The prodigious contrapuntal combinations so typical of the tonal works lose ground to relatively simple textures in which one or two salient lines predominate. But the rhythmic articulation of accompaniments fashioned out of serial forms in balanced rotation produces a wealth of motivic reference, as well as the play of rhythmic wit which is such a notable feature of Schoenberg's later scores. Thus the superimposition of ideas, with its risk of overloading, gives way to a finely graduated perspective in which the listener discovers with increasing familiarity ever more layers of meaning beyond the clearcut foreground, as his hearing travels towards the inaudible vanishing-point of total serial connection.

At the end of 1928 Schoenberg drafted the first version of the text of *Moses und Aron* (in the form of an oratorio) and composed the one-act comic opera *Von heute auf morgen*. The subjects of both works had been anticipated three years earlier in the two sets of short choral pieces opp.27 and 28. Most of these make considerable use of strict canonic or fugal writing, a feature that is taken up on a greatly expanded scale in the

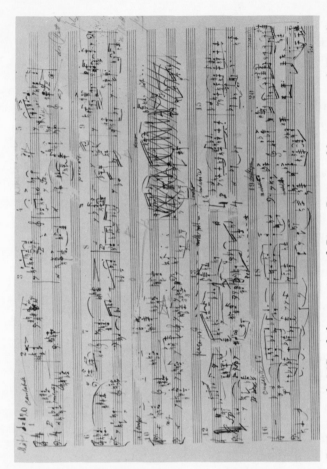

7. *Autograph MS of the opening of Piano Piece op.33a, composed December 1928–April 1929*

ensembles and choruses of the operas. The Three Satires op.28 deride the irresponsibility of modish modernity in music (especially Stravinsky's neo-classicism); *Von heute auf morgen* attacks the same thing in life. This is a comedy of marital strife and reconciliation involving a symmetrical quartet of characters: a wife brings her husband to heel when he takes an interest in an emancipated 'woman of today' by showing that she could play the same game if she wished. The little incident, which Gertrud Schoenberg with her husband's assistance turned into a very serviceable libretto, was suggested by the domestic life of the Schrekers – a source of frequent amusement to the Schoenbergs. The text makes its points bluntly, like most that Schoenberg had a hand in or wrote himself: his musical style is not primarily illustrative and prefers a simple basis for the wealth of comment and interpretation that it provides in its own terms. The opera adopts Classical procedures, but handles them rather freely. Recitative and arioso break into the set pieces, expanding them to accommodate great flexibility of pace and feeling as the bickering characters waver between good sense and self-indulgence. Schoenberg finds no broad comedy in the commonplace and absurd situations, but endless nuances of humour and sentiment which, no less than the extremes of spirituality and depravity in *Moses und Aron*, relate to perennial components in his expressive range.

It was another 18 months before Schoenberg finally began to compose *Moses und Aron*. In the meantime he produced several smaller works in which the relation to Classical form becomes looser. The first piano piece of op.33 and its slightly later companion (1931) each employ a pair of contrasting themes, but the first, at least, recalls the concentrated manner of op.23. At this time he became interested in the problem of film music. Unwilling to subordinate his music to the requirements of a real film he chose instead to illustrate in his *Begleitmusik zu einer Lichtspielszene* an imaginary and unfilmable sequence of emotions: threatening danger, fear, catastrophe. He employed a kind of free variation form, and thinned out his recent style considerably to suit the programmatic nature of the undertaking. Since 1916 Schoenberg had now and then used tonality in fragmentary sketches and occasional pieces (notably the beautiful *Weihnachtsmusik* of 1921 based on *Es ist ein' Ros' entsprungen*), but never in published works. In 1929, however, he made some folksong settings for a commission, and followed them up with two non-dodecaphonic male choruses, *Glück* and *Verbundenheit*, the second of which centres on D minor. Although the other four choruses that make up op.35 are dodecaphonic, the exceptions show that the urge to return to tonal composition was beginning to gain ground.

Moses und Aron, composed between 1930 and 1932, is Schoenberg's second great profession of faith, a sequel to *Die Jakobsleiter* dealing with the predicament of the chosen one in carrying out his prophetic task. Unlike the oratorio, however, the work is in no real sense unfinished, even though the short third act was never set to music. The reason for this lies in the subject itself. At the beginning of Act 1 God, speaking from the burning bush, assigns to Moses the role of prophet. Schoenberg had summed up the problems of revelation without distortion in the second chorus from op.27: 'You shall not make an image. For an image confines, limits, grasps what should remain limitless and unimaginable. An image demands a name which you can take only from what is little. You shall not worship the little! You must believe in the spirit, directly, without emotion, selflessly'. Moses complains that he lacks eloquence to express what he understands of God, who accordingly appoints Aaron as his spokesman. Aaron comes to meet Moses; he echoes Moses's thoughts in less uncompromising terms, and this is underlined by the casting of Moses as a speaker and Aaron as a lyrical tenor. They return together to bring the demoralized but expectant Israelites news of the new god who is to deliver them from Egyptian bondage. Moses tells them flatly that the one almighty, invisible and unimaginable God requires no sacrifices of them but complete devotion, and meets with a derision that Aaron can quell only by performing a series of three miracles, thereby substituting an image for the truth.

In Act 2 Aaron is obliged to still the people's doubt when Moses is away praying on the mountain by setting up a real image for them to worship in the form of the golden calf. The healing benefits of a faith so shallowly grounded are soon swept away by an orgy culminating in human sacrifice, suicide, lust and wholesale destruction. When Moses returns the calf vanishes at his word, but Aaron is able to defend his actions by pointing out that he is Moses's interpreter, and not an independent agent. The people are seen following yet another image, this time the pillar of fire, and Moses is left in despair. The uncomposed third act consists of another exchange between the brothers. This time Moses prevails. Aaron, who has been under arrest, is freed but falls dead; with all barriers to spiritual understanding removed the people will at length achieve unity with God. Schoenberg once suggested that Beethoven, Bruckner and Mahler had not been permitted to compose tenth symphonies because they might have revealed something that we are not permitted to know; a ninth seemed to represent a limit beyond which the composer must pass into the hereafter. To have composed music adequate to the idea of unity with God would have been to write a tenth symphony. At some level Schoenberg must have felt this from the outset, for the first two acts of the opera are dramatically and musically complete in themselves. But to remain true to his mission he could not admit that: it was his duty to continue to strive towards the expression of the inexpressible. To the end of his life he still spoke of finishing the work.

In its formal procedures *Moses und Aron* follows *Von heute auf morgen* in striking a balance between Classical number opera and Wagner's continuous symphonic manner, but on a far larger scale incorporating very big choral or orchestral movements. Schoenberg draws on every aspect of his music of the previous decade, and in the partly spoken texture of much of the choral writing looks back further. It is in every way his most comprehensive masterpiece, encompassing the stillness of the purely spiritual glimpsed momentarily in the opening bars, Moses's bitterness and resignation, Aaron's ecstatic eloquence and occasional weakness, and the people's jubilation, instability, mockery, violence and outright savagery. And it is noteworthy that the music interprets the stern morality of the libretto with a breadth of sympathy lacking in the neutral words.

That Schoenberg should now have sought relaxation in a less monumental task is not so surprising as his choice, which took the form of a pair of concertos for cello and for string quartet, based respectively on a keyboard concerto by M. G. Monn for which he had provided a continuo part some 20 years earlier, and Handel's Concerto grosso op.6 no.7 (the only one of the set that lacks separate concertino parts throughout). These works are often mistakenly classified as arrangements. However, whereas in his orchestrations of Bach and Brahms Schoenberg added nothing substantial to the original and never overstepped the style, the concertos are new compositions to almost the same degree as a set of variations on another composer's theme. Thus in each movement of the Cello Concerto he overlaid Monn's exposition with additional counterpoints and harmonies reaching as far forward as Brahms, or even later, and then continued independently in the same style. In the Quartet Concerto he preserved the complete outline of the original first movement and scarcely changed the second; on the other hand he radically recomposed the two remaining movements, taking only a few phrases from Handel in the third. In 1934 he crowned this group of works with a Suite in G for string orchestra in a similar style but based entirely on his own material. By way of indicating their secondary status he did not confer an opus number on any of them, yet they are brilliant compositions that he could certainly not have written earlier. The pressures towards the dissolution of tonality that haunt his older tonal works are entirely absent; the late works accept their terms of reference, and the clarity with which their abundant invention is projected derives directly from the serial works of the previous decade.

The one aspect of Schoenberg's serial music for which *Moses und Aron* had given only restricted opportunity was its abstract symphonic thought. This now became his chief concern again. After his tonal excursion he composed in 1935–6 the Violin Concerto and the Fourth String Quartet, his first 12-note works (apart from the three songs of op.48) since the opera, and with it the culminating productions of this period of his work. They are cast in the respective three- and four-movement moulds traditional in such works, but the individual movements abandon strict Classical layout. The first-movement recapitulations no longer correspond to the measure of the expositions, but are engulfed in the development, which continues unchecked to the close. The forward urge that marks all Schoenberg's music asserts itself so forcefully here that a return to single-movement structure through the breakdown of the divisions between movements might have been foretold. Such a return did indeed take place, but the transition was not a straightforward one.

IV Later works

Schoenberg's music was once more reaching a turning-point, even if a less acute one than those reached in 1908 or 1920. Since the latter date his progressive re-interpretation of earlier musical principles had led him from Baroque and Classical models to a more fluid formal approach analogous to that of the later 19th century. The next step could only bring him to his own work – not merely to single-movement form but beyond that to the achievement of his expressionist years. But this very achievement was to a considerable degree the basis for his reinterpretation. As though to understand his situation better he took up, in 1939, the sketches for the Second Chamber Symphony, begun in 1906 on the threshold of the crisis that now, though in a different way, confronted him for the second time. He completed the work in two movements, adding the last 20 bars of the first and about half of the second (from bar 309), but he also rescored and revised the remainder. In the process he increased the emphasis on a technical trait already prominent in the contemporary songs opp.12 and 14. Whereas the harmony of the First Chamber Symphony had been characterized by an abundance of complex suspensions and appoggiaturas, that of the Second tends to progress by stepwise movement in all parts. Schoenberg combined this technique with frequent 4th chords and similar combinations to very austere effect; indeed, the final coda strikes an unequivocally tragic note such as his later style would scarcely have countenanced. Perhaps recognition of this possibility in the material was an additional reason for his returning to it at this difficult period of his life.

In the previous year he had composed a setting of the *Kol nidre* in a tonal style which he hoped would prove acceptable in the synagogue. However, the work was found unsuitable for liturgical use because he had added an introduction and altered the traditional text in an attempt to strengthen its spiritual content. In order to give the main declaration of repentance and dedication 'the dignity of law', in his own phrase, he set it in march-like fashion and reinforced the effect with a harmonic severity that anticipates, in simpler terms, that of the chamber symphony. After finishing the symphony he still felt that his harmonic style just before his first pantonal works offered unused possibilities. He set about exploring them further in the D minor Variations on a Recitative for organ. Here, as in several pieces of the earlier time, harmonic complexity is controlled by unremitting reference to the tonic; there are also, however, serial features. In many ways this work and the next, the setting of Byron's *Ode to Napoleon*, form a complementary pair of opposites. Each is rooted in a special harmonic procedure that gives it a peculiarly individual sound. The D minor work borrows from serialism; the dodecaphonic *Ode* ends in E♭. The Variations respect the integrity of their melodic theme; the series of the *Ode* is freely permuted. The sequence of extraordinarily heterogeneous works starting with the *Kol nidre*, each employing a different technique for a particular end, shows the composer once again moving as though inadvertently towards a definite point, in this case the resumption of serial composition from a rather different angle.

The Piano Concerto of 1942 consists of one movement, less a conflation of several movements like the First Quartet and First Chamber Symphony than an expansion of a single sonata movement to embrace four symphonic characters in traditional sequence. As in the serial works up to 1936 all essential elements derive from the unpermuted series, but there are also strong affinities with the *Ode to Napoleon*. In the first place the music shares to some extent the quasi-tonal leanings of the *Ode*. This leads to more stable textures than are common in the earlier serial works, let alone the expressionist ones, and to symmetrical formal schemes, at least in outline. None of this suggests that Schoenberg was more closely engaged with the crisis of 1908 than he had been in 1936 – rather the reverse. But another legacy from the *Ode* changes the picture. At the very opening the serial melody is supported by free permutations of itself, its unusual tonal stability achieved through an unstable element. The consequences emerge later: chaos lies in wait at transitional points, above all at the end of each of the middle sections, where the chromatic totality becomes an undifferentiated stack of 4ths which momentarily endanger the work's identity. The abyss had opened before in Schoenberg's music, for instance in *Erwartung*, but never beneath so serene a surface. The effect is correspondingly disturbing.

The following year at the request of his publishers he composed a set of variations in G minor for band. It was intended for wide circulation and so couched in a straightforward tonal idiom, like the G major Suite for Strings, which had been written for college orchestras. It has all the vigour and ebullience of the earlier work, and as there his personality marks every bar no less firmly than in his more dissonant style. After this he wrote nothing for two years owing to deterioration in his health. When he resumed composing approximately one work a year, as he had done fairly regularly since his arrival in America, his bad eyesight obliged him to restrict his scale of activity. He wrote the first work of this last group in response to another commission. He was asked to compose the prelude to a suite for chorus and orchestra by various composers based on selections from the book of *Genesis*. Schoenberg evidently thought of God as creating the world out of divine order rather than primordial chaos, for the core of his compact piece consists of an eight-part double canon followed by two strettos that draw into their orbit the more amorphous elements from the opening. In the works of this last phase (except, of course, the folksong settings op.49), the tonal influence that had still been perceptible in the Piano Concerto recedes, and the language moves somewhat closer to that of the serial works up to 1936.

The longest and most wide-ranging of these late works is the String Trio of 1946. It is cast in a single movement expanded from within by the pressure of continuous and multifarious development. The different musical characters do not group themselves into clear subsidiary sections, as in the Piano Concerto, but alternate with a degree and frequency of contrast that Schoenberg had avoided since his expressionist period. Indeed, with this work he finally overtook his own earlier achievement and absorbed it into his later mode of thought. He divided the score into three 'parts' separated by two 'episodes' of different serial construction. The first part and episode correspond to an exposition, and the second part and episode to a development; the third part contains a truncated but unusually exact recapitulation and a coda. The structure recalls the first movement of the Fourth Quartet in the return early in the development to the codetta of the exposition, before the emergence of an important new melody.

The outpouring of elusive, visionary music held within this framework arose directly from the special circumstances of composition: Schoenberg had just recovered from an almost fatal heart attack, and he confessed that the experience was reflected in the Trio. It is not difficult to guess the direction of his thoughts. Having stood nearer than ever before to the truths that lay beyond man's reach in this world he was under the obligation to reveal what he could. The reinterpretation of expressionism towards which he had been moving suddenly took on a new urgency. Just as, nearly 40 years before, the attempt to lift all constraints from intuition had led to him placing his art in the service of faith, and eventually to a new order in composition, so he might now, from his present level, reach further still. But if the work contains intimations of the hereafter it is also concerned with this world: the melody heard in the second part and again in the coda recalls the music for the woman healed by faith, even though faith in an image, in Act 2 of *Moses und Aron*, and would seem to refer to his precarious recovery. There can be little doubt that the work was intended as a personal and spiritual testament, and it could have closed his life-work worthily.

In the event Schoenberg lived another five years and was able to compose a second testament in 1950. His first work in the interim, *A Survivor from Warsaw*, was wrung from him by a report of an occasion when Jews on their way to the gas chamber found courage in singing the *Shema Yisroel*, the command to love God, who is one lord. Though a short piece it made large demands. The orchestral accompaniment to the witness's spoken narration illustrates a reality more horrible than anything that Schoenberg could have imagined when he wrote his *Begleitmusik*, and his original melody for the Hebrew cantillation is an extraordinary conception, expressing a desperate tenacity that belongs very much to its author. The three folk-songs op.49 are new settings of tunes that he had already arranged in 1929. The two choral settings of *Es gingen zwei Gespielen gut*, the most elaborate in their respective sets, show a revealing shift of emphasis: the 1929 version takes the form of a complex set of canonic variations, that of 1948 is less intricate but allows the original melody to dissolve in the general texture of variation. Schoenberg's last instrumental work is entitled 'Phantasy for violin with piano accompaniment'. The description is exact: the violin part leads throughout, having even been written separately before the accompaniment. Melody accordingly dominates, limiting a tendency towards the sharp contrasts characteristic of the String Trio and checking their more disruptive consequences. This is the key to the work's special quality. It stands close to the Trio in many points of style, including melodic style, and in its subtlety of thematic continuity, but finds more consistent tranquillity.

The three religious choruses for mixed voices of op.50 were conceived at different times for different purposes and have little in common. However, at a time when he still hoped to finish the last one Schoenberg looked forward to their performance as a group. *Dreimal tausend Jahre* is a four-part setting of a short poem looking forward to God's return among the faithful in the new Israel. The close-knit textures and full harmony ally it to the male choruses op.35, whereas the mixture of singing and speech in the more dramatic six-part *De profundis* recalls *Moses und Aron* and throws into relief the varied soloistic phrases expressing repentance and supplication. The third, unfinished piece employs a speaker and an orchestra with the chorus. The text is a meditation on prayer by the composer himself, the first of the series of 'modern psalms' that occupied the last months of his life. Towards the end it speaks of the feeling of unity with God experienced in prayer. The passage is first given to the speaker, and should then have been taken up by the chorus. But at this point the composition breaks off, for it presented Schoenberg with the same task, at once impossible to fulfil yet central to his beliefs, as the third act of *Moses und Aron*: that of revealing through his music what it is not given to man to know. Although he still entertained the notion of working on *Die Jakobsleiter* and *Moses* until shortly before his death he must really have known that it was out of the question, and that the withdrawal into silence manifested in the *Modern Psalm* represented his final testament.

Since the time of his death Schoenberg's cardinal importance as an innovator has been very widely recognized. As a result most of his works are now assured of at least an occasional hearing. Yet although his idiom is no longer unfamiliar in a general sense, his music remains less easily accessible than that of his eminent pupils and contemporaries. One difficulty has been that musicians who shared his background and artistic assumptions, and might in principle have built up a tradition of performance – men such as Furtwängler, Walter, Kleiber and Klemperer, all of whom worked in Berlin when Schoenberg was there – failed to keep abreast of his development, while the more objective, uncommitted approach cultivated in recent years overlooks too much. But if the scarcity of good performances has not helped to dispel the wider public's indifference, neither does it entirely account for it. There would appear to be more fundamental causes that affect specialist audiences as well.

In 1930 Berg drew attention to the close parallel between Schoenberg's historical position and that of

Bach. He showed that a few small changes could make the assessment of the latter in Riemann's encyclopedia apply equally well to Schoenberg, who, like Bach, lived at a time of transition between two musical styles and succeeded in reconciling their opposing characteristics through his genius. Berg did not live to see his comparison further borne out by changes in taste after his teacher's death. Just as Bach's music held no interest for a generation preoccupied with the simpler language of early symphonic music, so the greater part of Schoenberg's work has had limited appeal for ears attuned to the broader effects of new sound resources and aleatory procedures. Its Bach-like density, proliferation and order run counter to the spirit of the age, making exceptional demands on the interpretative discipline of the performer and the sensibility of the listener. In the long run, however, these very qualities are likely to tell no less powerfully in its favour. Perhaps no other composer of the time has so much to offer.

WORKS

Edition: *A. Schoenberg. Sämtliche Werke* (Mainz, 1966–) [S]

Only a selection of the more considerable of Schoenberg's numerous unfinished compositions is included here. Many more are listed in Rufer (1959), and those up to 1933 are catalogued in greater detail in Maegaard, i (1972). Some fragments are published in Maegaard, iii (1972) [M]. Works without opus numbers are unpublished unless otherwise stated. For more precise details of composition dates see Rufer (1959) and Maegaard (1972).

Numbers in the right-hand column denote references in text.

OPERAS

op.		
17	Erwartung (monodrama, 1, M. Pappenheim), Aug–Sept 1909; Prague, Neues Deutsches Theater, 6 June 1924; vocal score by Schoenberg	8, 12, 41, 42
18	Die glückliche Hand (drama with music, 1, Schoenberg), 1910–Nov 1913; Vienna, Volksoper, 14 Oct 1924	10, 12, 42, 44, 46
32	Von heute auf morgen (opera, 1, M. Blonda [pseud. of G. Schoenberg]), Oct 1928–Jan 1929; Frankfurt, Opernhaus, 1 Feb 1930; vocal score by Schoenberg; S A/7	14, 53, 54
—	Moses und Aron (opera, 3, Schoenberg), May 1930–March 1932, Act 3 not composed; Der Tanz um das goldene Kalb perf. Darmstadt, 2 July 1951; Acts 1–2 perf. in concert, Hamburg, 12 March 1954; Acts 1–2 staged, Zurich, Stadttheater, 6 June 1957; (1957); S A/8	14, 23, 24, 48, 53, 54, 56, 65, 66, 67

(fragments)

—	Und Pippa tanzt (G. Hauptmann), Aug 1906–March 1907; prelude and recitative, short score, 68 bars	

CHORAL

—	Ei du Lütte (partsong, K. Groth), early; S A/18	
—	Friedlicher Abend senkt sich aufs Gefilde (partsong in canon, O. Kernstock), early; S A/18	
—	Viel tausend Blümlein auf der Au, partsong, early	5, 8, 9, 31, 32, 33, 44
—	Gurrelieder (J. P. Jacobsen, trans. R. F. Arnold), solo vv, choruses, orch, March 1900–March 1901, orchd Aug 1901–1903, July 1910–Nov 1911; (1912)	
13	Friede auf Erden (C. F. Meyer), SSAATTBB, insts ad lib, Feb–March 1907, acc. Oct 1911; S A/18	37
—	Der deutsche Michel (O. Kernstock), male vv, 1914 or 1915	
27	Four Pieces, SATB: Unentrinnbar (Schoenberg), Sept 1925; Du sollst nicht, du musst (Schoenberg), Oct 1925; Mond und Menschen (Tschan-Jo-Su, trans. Bethge), Oct 1925; Der Wunsch des Liebhabers (Hung-So-Fan, trans. Bethge), with cl, mand, vn, vc, Nov 1925; S A/18	53, 56
28	Three Satires (Schoenberg), SATB: Am Scheideweg, Nov 1925; Vielseitigkeit, Nov–Dec 1925; Der neue Klassizismus, with va, vc, pf, Nov–Dec 1925; pubd with appendix of three canons (see Canons below); S A/18	53, 54
—	Three Folksongs, SATB, Jan 1929 (1930): Es gingen zwei Gespielen gut; Herzlieblich Lieb, durch Scheiden; Schein uns, du liebe Sonne; S A/18	56, 65
35	Six Pieces (Schoenberg), male vv: Hemmung, Feb 1930; Gesetz, March 1930; Ausdrucksweise, March 1930; Glück, March 1929; Landsknechte, March 1930; Verbundenheit, April 1929; S A/18	56, 66
39	Kol nidre (Jewish liturgy in Eng. with alterations and introduction), speaker, chorus, orch, Aug–Sept 1938; S A/19	16, 61
44	Prelude 'Genesis' (textless), SATB, orch, Sept 1945; S A/19	63
46	A Survivor from Warsaw (Schoenberg), narrator, male vv, orch, Aug 1947; S A/19	65
49	Three Folksongs, SATB, June 1948: Es gingen zwei Gespielen gut (Two comely maidens); Der Mai tritt ein mit Freuden (Now May has come with gladness); Mein Herz in steten Treuen (To her I shall be faithful); S A/19	64, 65
50a	Dreimal tausend Jahre (D. D. Runes), SATB, April 1949; S A/19	66
50b	De profundis (Ps cxxx in Heb.), SSATBB, June–July 1950; S A/19	66
50c	Modern Psalm (Der erste Psalm) (Schoenberg), speaker, chorus, orch, Oct 1950, inc; S A/19	66

(fragments)

—	Wenn weder Mond noch Stern am Himmel stehn (L. Pfau), male vv, wind ens, June 1897; 54 bars	
—	Darthulas Grabgesang (Goethe), 14vv, orch, April 1903; vocal score, 65 bars	

Symphony with choral movements, 1914–15; sketches; M (ex-tracts) — 45

Die Jakobsleiter (oratorio, Schoenberg), solo vv, choruses, orch, June 1917–July 1922, rev. begun Oct 1944 and abandoned after bar 104: first half only composed 1917 and in *A. Schönberg: Texte* (Vienna, 1926); vocal score, arr. W. Zillig (1975) — 11, 23, 24, 25, 48, 50, 67

Israel Exists Again (Schoenberg), chorus, orch, March–June 1949; short score, 55 bars; S A/19

ORCHESTRAL

— Adagio, harp, str, early

— Gavotte and Musette (in Olden Style), str, March 1897

4 Verklärte Nacht, arr. str orch 1917, 2nd version 1943 — 5, 7, 9, 21

5 Pelleas und Melisande, sym. poem, after Maeterlinck, July 1902–Feb 1903

9 Chamber Symphony no.1, arr. full orch Nov 1922, 2nd version April 1935; S A/12

10 String Quartet no.2, arr. S, str orch ?1919

16 Five Orchestral Pieces: no.1 May 1909, nos.2–3 June 1909, no.4 July 1909, no.5 Aug 1909; S A/12; arr. reduced orch, Sept 1949 — 8, 9, 40, 41

31 Variations for Orchestra, May 1926, July–Aug 1928 — 14, 49, 52

34 Begleitmusik zu einer Lichtspielszene, Oct 1929–Feb 1930 — 14, 56, 65

— Cello Concerto [after Monn: Clavicembalo Concerto in D, 1746], Nov 1932–Jan 1933 (1935); S A/27: red. by Schoenberg for vc, pf; S B/27 — 14, 59

— String Quartet Concerto [after Handel: Concerto grosso op.6 no.7], May–Aug 1933 (1963); S A/27 — 15, 59

36 Suite, G, str, Sept–Dec 1934 (1935)

36 Violin Concerto, 1935–Sept 1936; S A/15 — 16, 59, 63

38 Chamber Symphony no.2, Aug 1906–Dec 1916, Aug–Oct 1939; S A,B/11 — 16, 60

42 Piano Concerto, July–Dec 1942; S A/15 — 16, 37, 61

43a Theme and Variations, band, completed July 1943; arr. orch as op.43b, summer 1943 — 62

(fragments)

— Waltz, str, early; 10 sections completed

— Serenade, small orch, 1896; 1st movt completed, 2nd and 3rd inc.

— Frühlings Tod, sym. poem, after Lenau, 1898; 254 bars of which 136 fully scored

— Symphony, G, Feb 1900; Introduction, g, pf score, 73 bars — 63

Passacaglia, March 1920; sketches; M

Symphony, Jan–Feb 1937; short score, 30–50 bars of each of the 4 movts — 48

— untitled work, Oct–Nov 1946; short score, 28 bars

— untitled work, April 1948; short score, 25 bars

CHAMBER

— 'Alliance' Walzer, 2 vn, early

— 'Sonnenschein' Polka schnell, 2 vn, early

— 3 Songs without Words, 2 vn, early

— untitled work, d, vn, pf, early

— Presto, C, str qt, early

— String Quartet, D, summer–autumn 1897 (1966) — 3, 28, 29

— Scherzo in F and Trio in a, str qt, July 1897; presumably rejected 2nd movt of preceding

4 Verklärte Nacht, after Dehmel, 2 vn, 2 va, 2 vc, completed Dec 1899 — 3, 6, 18, 30, 31

7 String Quartet no.1, d, summer 1904–Sept 1905 — 7, 35

9 Chamber Symphony no.1, 15 insts, completed July 1906; S A,B/11 — 7, 36, 51, 61

10 String Quartet no.2, with S in movts 3 'Litanei' and 4 'Entrückung' (George), March 1907–Aug 1908 — 7, 37, 39

— Three Pieces, wind qnt, org/harmonium, cel, str qt, db, Feb 1910, no.3 inc. (?1970) — 41

24 Die eiserne Brigade, march, pf qnt, 1916 (1978)

— Serenade, cl, b cl, mand, gui, vn, va, vc, with B in movt 4 'O könnt ich je der Rach' an ihr genesen' (Petrarch, trans. K. Förster), Aug 1920–April 1923 — 12, 49, 50

26 Weihnachtsmusik, 2 vn, vc, harmonium, pf, Dec 1921 (1975) — 56

26 Wind Quintet, April 1923–Aug 1924 — 12, 49, 51

29 Suite, Eb-cl/fl, cl, b cl/bn, pf, vn, va, vc, Jan 1925–May 1926 — 13, 51

30 String Quartet no.3, Jan–March 1927; S A/21 — 14, 52

37 String Quartet no.4, April–July 1936; S A/21 — 16, 60

45 String Trio, Aug–Sep 1946; S A/21 — 17, 64

47 Phantasy, vn, pf, March 1949 — 66

(fragments)

— String Quartet, C, early; 41 bars

— Clarinet Quintet, d; 28 bars

— Toter Winkel, after G. Falke, 2 vn, 2 va, 2 vc, ?before op.4; 31 bars

— Chamber Symphony, a, ?before op.9; 22 bars

— Fugue, d, str qt, March 1904; 80 bars

— String Quintet, D, winter 1904–5; 22 bars

— Ein Stelldichein, after Dehmel, ob, cl, pf, vn, vc, Oct 1905; 90 bars (1981)

— String Septet, March 1918; 25 bars

— Tempo zwischen langsamem Walzer und Polacca, movt intended for op.24, Aug 1920; 40 bars; M

— Gerpa, F, for Schoenberg's son Georg (hn + pf) and himself (vn + pf + harmonium), Nov 1922; theme and 3 variations completed

— Sonata, vn, pf, Jan–Feb 1928; 43 bars

— String Quartet, June 1949; openings of all 4 movts, 36 complete bars in all

SOLO VOCAL

(for 1 v, pf unless otherwise stated)

— Songs, almost all before 1900; Dass gestern eine Wespe Dich; Dass schon die Maienzeit vorüber (A. Christen); Der Pflanze, die dort über dem Abgrund (Pfau); Drüben geht die Sonne scheiden (Schilflied) (Lenau); Du kehrst mir den Rücken (Pfau); Du musst nicht meinen (Mannesbangen) (Dehmel); Duftreich ist die Erde (Ecloge) (W …) (1975); Einsam bin ich und alleine (Pfau); Einst hat vor deines Vaters Haus; Es ist ein Flüstern in der Nacht, T, str qt; Es steht sein Bild noch immer da (Gedenken), S A/1; Gott grüss dich Marie (Pfau); Ich grüne wie die Weide grünt (W. Wackernagel); Ich hab' zum Brunnen ein Krüglein gebracht (Das zerbrochene Krüglein) (M. Greif); Im Fliederbusch ein Vöglein sass (R. Reinick); Juble, schöne junge Rose; Klein Vögelein, du zwitscherst fein; Könnt' ich je zu dir mein Licht (Pfau); Lass deine Sichel rauschen (Lied der Schnitterin) (Pfau); Mädel, lass das Stricken (Nicht doch!) (Dehmel); Mein Herz das ist ein tiefer Schacht; Mein Schatz ist wie ein Schneck (Pfau); Nur das thut mir so bitter weh (O. von Redwitz); Sang ein Bettlerpärlein am Schenkentor (Mädchenlied) (P. Heyse); Waldesnacht, du wunderkühle (Heyse); Warum bist Du aufgewacht

— In hellen Träumen hab ich Dich oft geschaut (A. Gold), 1893

— Du kleine bist so lieb und hold (Zweifler) (Pfau), ?1895

— War ein Blümlein wunderfein (Vergissmeinnicht) (Pfau), ?1895

— In meinem Garten die Nelken (Mädchenlied) (E. Geibel), ?1896

— Als mein Auge sie fand (Sehnsucht) (J. C. von Zedlitz), ?1896

— Aprilwind, alle Knospen (Mädchenfrühling) (Dehmel), Sept 1897

1 Two Songs (K. von Levetzow), Bar, pf, ?1898: Dank; Abschied; S A/1   3, 30

— Sie trug den Becher in der Hand (Die Beiden) (Hofmannsthal), April 1899

2 Four Songs: Erwartung (Dehmel), Aug 1899; Schenk mir deinen goldenen Kamm (Dehmel), ?1899, Erhebung (Dehmel), Nov 1899; Waldsonne (J. Schlaf); S A/1   3, 30, 34

— Dunkelnd über den See (Gruss in die Ferne) (H. Lingg), Aug 1900 (1976)

— Lied der Waldtaube [from Gurrelieder], arr. Mez, 17 insts. 1900, arr. Dec 1922 (1923); S A/3

— Cabaret songs: Der genügsame Liebhaber (H. Salus), April 1901 (1975); Einfältiges Lied (Salus), April 1901 (1975); Nachtwandler (G. Falke), S, pic, F-tpt, side drum, pf, April 1901 (1969); Jedem das Seine (Colly), June 1901 (1975); Mahnung (G. Hochstetter), July 1901 (1975); Galathea (Wedekind), Sept 1901 (1975); Gigerlette (O. Bierbaum) (1975); Seit ich so viele Weiber sah (Aus dem Spiegel von Arcadia) (Schikaneder) (1975)   5

— Deinem Blick mich zu bequemen (Goethe), Jan 1903

3 Six Songs, Mez/Bar, pf: Wie Georg von Frundsberg von sich selber sang (Des Knaben Wunderhorn), March 1903; Die Aufgeregten (G. Keller), Nov 1903; Warnung (Dehmel), May 1899; Hochzeitslied (Jacobsen, trans. Arnold), ?1900; Geübtes Herz (Keller), Sept–Nov 1903; Freihold (Lingg), Nov 1900; S A/1   3, 30, 31, 33

6 Eight Songs: Traumleben (J. Hart), Dec 1903; Alles (Dehmel), Sept 1905; Mädchenlied (P. Remer), Oct 1905; Verlassen (H. Conradi), Dec 1903; Ghasel (Keller), Jan 1904; Am Wegrand (J. H. Mackay), Oct 1905; Lockung (K. Aram), Oct 1905; Der Wanderer (Nietzsche), ?April–Oct 1905; S A/1   33, 34, 38

8 Six Orchestral Songs: Natur (H. Hart), Dec 1903–March 1904; Das Wappenschild (Des Knaben Wunderhorn), Nov 1903–May 1904; Sehnsucht (Des Knaben Wunderhorn), completed April 1905; Nie ward ich, Herrin, müd' (Petrarch, trans. Förster), June–July 1904; Voll jener Süsse (Petrarch, trans. Förster), completed Nov 1904; Wenn Vöglein klagen (Petrarch, trans. Förster), completed Nov 1904; S A/3   33, 34

12 Two Ballads, March–April 1907: Jane Grey (H. Ammann), Der verlorene Haufen (V. Klemperer); S A/1   37, 61

(all published in S A/18)

4-part canon 'O dass der Sinnen doch so viele sind!' (Goethe), ?April 1905

4-part canon 'Wenn der schwer Gedrückte klagt' (Goethe), ?April 1905

'Eyn doppelt Spiegel- und Schlüssel-Kanon', 4 parts, Feb 1922

'Ein Spruch und zwei Variationen über ihn: O glaubet nicht, was ihr nicht könnt, sei wertlos', op.28 App.1, 4 parts (Schoenberg), Dec 1925–Jan 1926

Canon for string quartet, op.28 App.2, Feb 1926

'Legitimation als Canon: Wer Ehr erweist, muss selbst davon besitzen', op.28 App.3, 6 parts (Schoenberg), April 1926

3-part canon for D. J. Bach 'Wer mit der Welt laufen will' (Schoenberg), March 1926, July 1934

4-part canon by augmentation and diminution, April 1926

4-part canon for Erwin Stein 'Von meinen Steinen' (Schoenberg), Dec 1926

'Arnold Schönberg beglückwünscht herzlichst Concert Gebouw', 5 parts (Schoenberg), March 1928

Canon in 3 keys for the Genossenschaft deutscher Tonsetzer, 5 parts, April 1928

Mirror canon for string quartet, April 1931

2-part mirror canon for Herrmann Abraham 'Spiegle Dich im Werk' (Schoenberg), Dec 1931

4-part mirror canon, Dec 1931

Mirror canon for string quartet, ?1931

4-part mirror canon for Carl Moll, Dec 1932

3-part puzzle canon for Carl Engel 'Jedem geht es so (No man can escape)' (Schoenberg in Ger. and Eng.), April 1933, text 1943

3-part puzzle canon for Carl Engel 'Mir auch ist es so ergangen (I, too, was not better off)' (Schoenberg in Ger. and Eng.), April 1933, text 1943

4-part perpetual canon, April 1933

4-part mirror canon, April 1933

4-part mirror canon, Dec 1933

3-part puzzle canon, March 1934

4-part puzzle canon by augmentation and diminution, March 1934

4-part puzzle canon, March 1934

4-part puzzle canon for Rudolph Ganz 'Es ist so dumm' (Schoenberg), Sept 1934

4-part mirror canon, Sept 1934

4-part mirror canon, 1934

7-part perpetual canon, 1934

4-part mirror canon, 1934

4-part perpetual canon with free bass for Alban Berg 'Darf ich eintreten' (Schoenberg), Feb 1935

4-part mirror canon for Frau Charlotte Dieterle, Nov 1935

4-part mirror canon, Jan 1936

4-part double canon, 1938

4-part canon 'Mr Saunders I owe you thanks' (Schoenberg), Dec 1939

4-part mirror canon, June 1943

4-part canon for Richard Rodzinsky 'I am almost sure, when your nurse will change your diapers' (Schoenberg), March 1945

4-part double canon for Thomas Mann on his 70th birthday, June 1945

4-part canon 'Gravitationszentrum eigenen Sonnensystems' (Schoenberg), Aug 1949

(fragments)

'Gutes thu rein aus des Guten Lieben' (Goethe), ?April 1905; coda inc.

'Dümmer ist nichts zu ertragen' (Goethe), ?April 1905; lacking coda

'Wer geboren in bös'sten Tagen' (Goethe), ?April 1905; lacking coda

ARRANGEMENTS

H. Susaneck: *Irmen Walzer*, 2 vn; R. Waldman: *So wie du*, 2 vn; *Wiener Fiakerlied*, 2 vn; all early

A. Zemlinsky: *Sarema*, vocal score, summer 1897

H. Schenker: *Vier syrische Tänze*, orchd 1903

J. S. Bach: Chorale Prelude *'Komm, Gott, Schöpfer, heiliger Geist'* BWV 631, orchd April 1922 (1925)   59

J. S. Bach: Chorale Prelude *'Schmücke dich, O liebe Seele'* BWV 654, orchd April–June 1922 (1925)   59

Johann Strauss (ii): *Kaiserwalzer* op.437, fl, cl, pf qnt, April 1925 (c 1960)

J. S. Bach: *Prelude and Fugue, Eb, BWV 552*, orchd May–Oct 1928 (1929)   59

J. Brahms: *Piano Quartet, g, op.25*, orchd May–Sept 1937; S A/26   59

Hack-work (in early years Schoenberg scored some 6000 pages of operettas by Zepler and others; the following examples of his hack-work, except for the second, were published): H. van Eyken: *Lied der Walküre* (F. Dahn), orchd ?1901; B. Zepler: *Mädchenreigen*, orchd April 1902; A. Lortzing: *Der Waffenschmied von Worms*, pf duet, ?1903; G. Rossini: *Il barbiere di Siviglia*, pf duet, ?1903; F. Schubert: *Rosamunde: overture, entr'actes and ballet*, pf duet, ?1903

Fragments: *Aberglaube*, opera lib, early, 2 acts and beginning of 3rd: *Odoaker*, opera lib, early, 3 opening scenes; *Die Schildbürger*, comic opera lib, after G. Schwab, June–July 1901, 2 of 3 acts

THEORETICAL AND PEDAGOGICAL

Harmonielehre, spring 1910–July 1911 (Vienna, 1911, rev. 3/1922; Eng. trans., abridged, 1948, complete, 1978) [8, 20, 21]

Models for Beginners in Composition, completed Sept 1942 (Los Angeles, 1942, enlarged 2/1943, rev. 3/1972 by L. Stein) [20, 28]

Structural Functions of Harmony, completed March 1948, ed. H. Searle (London, 1954, rev. 2/1969 by L. Stein) [20]

Preliminary Exercises in Counterpoint, 1936–50, ed. L. Stein (London, 1963) [20]

Fundamentals of Musical Composition, 1937–48, ed. L. Stein (London, 1967) [20, 28]

Fragments: *Das Komponieren mit selbständigen Stimmen*, June 1911; *Die Lehre von Kontrapunkt*, Oct 1926; *Der musikalische Gedanke und seine Darstellung*, 1925–36, 3 drafts under similar titles

ESSAYS, LETTERS ETC

Style and Idea (New York, 1950) [15 essays]

Briefe, selected and ed. E. Stein (Mainz, 1958; Eng. trans., enlarged, 1964) [17]

Schöpferische Konfessionen, ed. W. Reich (Zurich, 1964)

Testi poetici e drammatici, ed. L. Rognoni (Milan, 1967)

Arnold Schönberg—Franz Schreker: Briefwechsel, ed. F. C. Heller (Tutzing, 1974)

Berliner Tagebuch, ed. J. Rufer (Frankfurt, 1974)

F. Busoni: *Entwurf einer neuen Aesthetik der Tonkunst, mit handschrift-lichen Anmerkungen von A. Schönberg* (Frankfurt, 1974)

Style and Idea, ed. L. Stein (London, 1975) [104 essays]

Gesammelte Schriften, i, ed. I. Vojtěch (Frankfurt, 1976)

The Arnold Schoenberg-Hans Nachod Collection, ed. J. Kimmey (Detroit, 1979) [Letters and early compositions]

Arnold Schönberg, Wassily Kandinsky: Briefe, Bilder und Dokumente einer aussergewöhnlichen Begegnung, ed. J. Hahl-Koch (Salzburg, 1980)

For list of unpubd writings see Rufer (1959); for bibliography of pubd writings see Brinkmann: *Arnold Schönberg: Drei Klavierstücke Op. 11* (Wiesbaden, 1969)

Continuo realizations, 1911 or 1912: M. G. Monn: *Sinfonia a 4, A* (1912); M. G. Monn: *Vc Conc...g* (1912) also arr. vc, pf (1913) and cadenzas, S B 27; M. G. Monn: *Cembalo Conc.. D* (1912); J. C. Monn: *Divertimento, D* (1912); F. Tůma: *Sinfonia a 4, e* (1968); F. Tůma: *Partita a 3, A* (1968); F. Tůma: *Partita a 3, c* (1968); F. Tůma: *Partita a 3, G* (1968)

Songs orchd for Julia Culp: L. van Beethoven: *Adelaide* op.46, Feb 1912; C. Loewe: *Der Nöck* op.129 no.2, autumn 1912; F. Schubert: *Three songs*, Sept 1912

Arrs. for the Society for Private Musical Performances (Schoenberg had a hand in various reductions for ensemble of his own and other works, but very few are wholly his; see L. Stein (1966)): Johann Strauss (ii): *Rosen aus dem Süden* op.388, harmonium, pf qnt, May 1921; Johann Strauss (ii): *Lagunenwalzer* op.411, harmonium, pf qnt, May 1921 (the arr. of Busoni's *Berceuse élégiaque* is not by Schoenberg)

Instrumentation exercises for teaching purposes, summer 1921: F. Schubert: *Ständchen* D889, 1v, cl, bn, mand, gui, str qt; L. Denza: *Funiculì, funiculà*, cl, gui, mand, str trio; J. Sioly: *Weil i a alter Drahrer bin*, cl, gui, mand, str trio

Principal publishers: Belmont, Dreililien, Hansen, G. Schirmer, Schott, Universal

MSS in University of Southern California School of Music, Los Angeles (composer's collection); Library of Congress, Music Division, Washington DC; North Texas State University Music Library, Denton; Universal Edition, Vienna; Pierpont Morgan Library, New York (Robert Owen Lehman collection)

WRITINGS

TEXTS WITHOUT MUSIC

Totentanz der Prinzipien, Jan 1915 [for Sym. sketched 1914–15]; pubd in A. Schönberg: *Texte* (Vienna, 1926)

Wendepunkt, ? Dec 1916 [for Chamber Sym. no.2 as melodrama]; pubd in Maegaard, i (1972) [13]

Requiem, first section 1920 or 1921, rest Nov 1923; pubd in A. Schönberg: *Texte* (Vienna, 1926) [14, 23]

Der biblische Weg, drama, June 1926–July 1927 [18]

Psalmen, Gebete und andere Gespräche mit und über Gott, Sept 1950–July 1951 (Mainz, 1956) [16 pieces, orig. entitled 'Modern Psalms', the last inc., the first partly composed as op.50c]

BIBLIOGRAPHY
MONOGRAPHS

E. Wellesz: *Arnold Schönberg* (Vienna, 1921; Eng. trans., rev., 1925/*R*1971)

P. Stefan: *Arnold Schönberg: Wandlung, Legende, Erscheinung, Bedeutung* (Vienna, 1924)

H. Wind: *Die Endkrise der bürgerlichen Musik und die Rolle Arnold Schönbergs* (Vienna, 1935)

R. Leibowitz: *Schönberg et son école* (Paris, 1947; Eng. trans., 1949/*R*1975)

D. Newlin: *Bruckner, Mahler, Schoenberg* (New York, 1947; Ger. trans., 1954; rev. 2/1979)

R. Leibowitz: *Introduction à la musique de douze sons* (Paris, 1949)

H. Stuckenschmidt: *Arnold Schönberg* (Zurich, 1951, rev. 2/1957; Eng. trans., 1959)

J. Rufer: *Die Komposition mit zwölf Tönen* (Berlin, 1952; Eng. trans., 1954)

L. Rognoni: *Espressionismo e dodecafonia* (Turin, 1954, rev. 2/1966 as *La scuola musicale di Vienna*; Eng. trans., 1977)

J. Rufer: *Das Werk Arnold Schönbergs* (Kassel, 1959, rev. 2/1975; Eng. trans., rev., 1962)

A. Payne: *Schoenberg* (London, 1968)

W. Reich: *Arnold Schönberg oder der konservative Revolutionär* (Vienna, 1968; Eng. trans., 1971)

R. Leibowitz: *Schoenberg* (Paris, 1969)

J. Maegaard: *Studien zur Entwicklung des dodekaphonen Satzes bei Arnold Schönberg* (Copenhagen, 1972)

E. Freitag: *Arnold Schönberg in Selbstzeugnissen und Bilddokumenten* (Reinbek, 1973)

H. Stuckenschmidt: *Schönberg: Leben, Umwelt, Werk* (Zurich, 1974; Eng. trans., 1977)

G. Manzoni: *Arnold Schönberg: l'uomo, l'opera, i testi musicati* (Milan, 1975)

C. Rosen: *Arnold Schoenberg* (New York, 1975)

G. Schubert: *Schönbergs frühe Instrumentation* (Baden-Baden, 1975)

M. Macdonald: *Schoenberg* (London, 1976)

J. Maegaard: *Praeludier til musik af Schönberg* (Copenhagen, 1976)

K. Velten: *Schönbergs Instrumentation Bachscher und Brahmsscher Werke als Dokumente seines Traditions-verständnisses* (Regensburg, 1976)

M. Pfisterer: *Studien zur Kompositionstechnik in den frühen atonalen Werken von Arnold Schönberg* (Neuhausen–Stuttgart, 1978)

T. Satoh: *A Bibliographic Catalog with Discography and a Comprehensive Bibliography of Arnold Schoenberg* (Tokyo, 1978)

75

A. Lessem: *Music and Text in the Works of Arnold Schoenberg* (Ann Arbor, 1979)

U. Thieme: *Studien zum Jugendwerk Arnold Schönbergs: Einflüsse und Wandlungen* (Regensburg, 1979)

D. Newlin: *Schoenberg Remembered: Diaries and Recollections (1938–76)* (New York, 1980)

COLLECTIONS OF ARTICLES AND ESSAYS

Der Merker, ii/17 (1911) [Schoenberg issue]

Arnold Schönberg: mit Beiträgen von Alban Berg, Paris von Gutersloh [and others] (Munich, 1912)

Arnold Schönberg zum fünfzigsten Geburtstage (Vienna, 1924)

'Schönberg und seine Orchesterwerke', *Pult und Taktstock*, iv (1927), March–April

Arnold Schönberg zum 60. Geburtstag (Vienna, 1934)

M. Armitage, ed.: *Schoenberg* (New York, 1937)

Canon, iii/2 (1949) [Schoenberg issue]

Stimmen (1949), no.16 [Schoenberg issue]

The Score (1952), no.6 [Schoenberg issue]

E. Stein: *Orpheus in New Guises* (London, 1953)

A. Webern: *Der Weg zur neuen Musik* (Vienna, 1960; Eng. trans., 1963)

B. Boretz and E. Cone, eds.: *Perspectives on Schoenberg and Stravinsky* (Princeton, 1968)

'Towards the Schoenberg Centenary', *PNM*, xi–xiii (1972–5)

E. Hilmar, ed.: *Arnold Schönberg, Gedenkausstellung 1974* (Vienna, 1974)

I. Kongress der Internationalen Schönberg–Gesellschaft: Vienna 1974, ed. R. Stephan

ÖMz, xxix/6 (1974) [Schoenberg issue]

Zeitschrift für Musiktheorie, v/1 (1974) [*Schoenberg issue*]

Journal of the Arnold Schoenberg Institute (1976–)

Mf, xxix/4 (1976) [Schoenberg issue]

C. Dahlhaus: *Schönberg und andere: gesammelte Aufsätze zur neuen Musik* (Mainz, 1978)

Musik-Konzepte, Sonderband, Arnold Schönberg (Munich, 1980)

SEPARATE ARTICLES

A. Nadel: 'Arnold Schönberg', *Die Musik*, xi/3 (1912), 353

C. Somigli: 'Il modus operandi di Arnold Schönberg', *RMI*, xx (1913), 583

P. Bekker: 'Schönberg', *Melos*, ii (1921), 123

L. Henry: 'Arnold Schönberg', *MO*, xliv (1921), 420, 511

C. Gray: 'Arnold Schönberg, a Critical Study', *ML*, iii (1922), 73

F. Wohlfahrt: 'Arnold Schönbergs Stellung innerhalb der heutigen Musik', *Die Musik*, xvi (1924), 894

Bibliography

R. van den Linden: 'Arnold Schönberg', *ML*, vii (1926), 322; viii (1927), 38

K. Westphal: 'Schönbergs Weg zur Zwölfton-Musik', *Die Musik*, xxi (1929), 491

A. Machabey: 'Schönberg', *Le ménestrel*, xcii (1930), 81, 245, 257

A. Weiss: 'The Lyceum of Schönberg', *MM*, ix (1932), 99

H. Gerigk: 'Eine Lanze für Schönberg', *Die Musik*, xxvii (1934), 87

D. J. Bach: 'A Note on Arnold Schoenberg', *MQ*, xxii (1936), 8

R. Hill: 'Schoenberg's Tone-rows and the Tonal System of the Future', *MQ*, xxii (1936), 14

H. Jalowetz: 'On the Spontaneity of Schoenberg's Music', *MQ*, xxx (1944), 385

D. Milhaud: 'To Arnold Schoenberg on his Seventieth Birthday: Personal Recollections', *MQ*, xxx (1944), 379

R. Sessions: 'Schoenberg in the United States', *Tempo* (1944), no.9, p.2; rev. in *Tempo* (1972), no.103, p.8

D. Newlin: 'Arnold Schoenberg's Debt to Mahler', *Chord and Dischord*, ii/5 (1948), 21

R. Leibowitz: 'Besuch bei Arnold Schönberg', *SMz*, lxxxix (1949), 324

D. Newlin: 'Schoenberg in America', *Music Survey*, i (1949), 128, 185

T. Wiesengrund-Adorno: 'Schönberg und der Fortschritt', *Philosophie der neuen Musik* (Tübingen, 1949; Eng. trans., 1973)

W. Rubsamen: 'Schoenberg in America', *MQ*, xxxvii (1951), 469

R. Vlad: 'L'Ultimo Schönberg', *RaM*, xxi (1951), 106

A. Duhamel: 'Arnold Schoenberg, la critique, et le monde musical contemporain', *ReM* (1952), no.212, p.77

H. Keller: 'Unpublished Schoenberg Letters: Early, Middle and Late', *Music Survey*, iv (1952), 499

G. Perle: 'Schoenberg's Later Style', *MR*, xiii (1952), 274

T. Wiesengrund-Adorno: 'Arnold Schönberg 1874–1951', *Neue Rundschau*, lxiv (1953), 80; repr. in T. Wiesengrund-Adorno: *Prismen* (Munich, 1963), 147

H. Eisler: 'Arnold Schönberg', *Sinn und Form*, vii (1955), 5

W. Reich: 'Alban Berg als Apologet Arnold Schönbergs', *SMz*, xcv (1955), 475

T. Wiesengrund-Adorno: 'Zum Verständnis Schönbergs', *Frankfurter Hefte*, x (1955), 418

J. Birke: 'Richard Dehmel und Arnold Schönberg, ein Briefwechsel', *Mf*, xi (1958), 279; xvii (1964), 60

'Letters of Webern and Schoenberg to Roberto Gerhard', *The Score* (1958), no.24, p.36

H. Stuckenschmidt: 'Stil und Ästhetik Schönbergs', *SMz*, xcviii (1958), 97

P. Gradenwitz: 'Schönbergs religiöse Werke', *Melos*, xxvi (1959), 330; Eng. trans. in *MR*, xxi (1960), 19

77

D. Kerner: 'Schönberg als Patient', *Melos*, xxvi (1959), 327

W. Reich: 'Ein unbekannter Brief von Arnold Schönberg an Alban Berg', *ÖMz*, xiv (1959), 10

H. Oesch: 'Hauer und Schönberg', *ÖMz*, xv (1960), 157

L. Rognoni: 'Gli scritti e i dipinti di Arnold Schönberg', *L'approdo musicale*, iii (1960), 95

F. Glück: 'Briefe von Arnold Schönberg an Adolf Loos', *ÖMz*, xvi (1961), 8

J. Maegaard: 'A Study in the Chronology of op.23–26 by Arnold Schoenberg', *DAM*, ii (1962), 93

G. Marbach: 'Schlemmers Begegnungen mit Schönberg, Scherchen und Hindemith', *NZM*, Jg.123 (1962), 530

N. Notowicz: 'Eisler und Schönberg', *DJbM*, viii (1963), 8

T. Wiesengrund-Adorno: 'Über einige Arbeiten Arnold Schönbergs', *Forum*, x (1963), 378, 434

R. Nelson: 'Schoenberg's Variation Seminar', *MQ*, l (1964), 141

F. Prieberg: 'Der junge Schönberg und seine Kritiker', *Melos*, xxxi (1964), 264

G. Schuller: 'A Conversation with Steuermann', *PNM*, iii (1964), 22

R. Steiner: 'Der unbekannte Schönberg: aus unveröffentlichten Briefen an Hans Nachod', *SMz*, civ (1964), 284

D. Dille: 'Die Beziehung zwischen Bartók und Schönberg', *Dokumenta bartókiana*, ii (1965), 53

I. Vojtěch: 'Arnold Schönberg, Anton Webern, Alban Berg: unbekannte Briefe an Erwin Schulhoff', *MMC*, xviii (1965), 31

P. Friedheim: 'Rhythmic Structure in Schoenberg's Atonal Compositions', *JAMS*, xix (1966), 59

V. Fuchs: 'Arnold Schönberg als Soldat im ersten Weltkrieg', *Melos*, xxxiii (1966), 178

R. Jung: 'Arnold Schönberg und das Liszt-Stipendium', *BMw*, viii (1966), 56

P. Odegard: 'Schönberg's Variations: an Addendum', *MR*, xxvii (1966), 102

L. Stein: 'The Privataufführungen revisited', *Paul A. Pisk: Essays in his Honor* (Austin, 1966), 203

'Unveröffentlichte Briefe an Alfredo Casella', *Melos*, xxxiv (1967), 45

D. Lewin: 'Inversional Balance as an Organizing Force in Schoenberg's Music and Thought', *PNM*, vi/2 (1968), 1

D. Newlin: 'The Schoenberg–Nachod Collection, a Preliminary Report', *MQ*, liv (1968), 31

E. Klemm: 'Der Briefwechsel zwischen Arnold Schönberg und dem Verlag C. F. Peters', *DJbM*, xv (1970), 5

H. Byrns: 'Meine Begegnung mit Arnold Schönberg', *Melos*, xxxviii (1971), 234

Bibliography

R. Vlad: 'Arnold Schönberg schreibt an Gian Francesco Malipiero', *Melos*, xxxviii (1971), 461

R. Stephan: 'Ein unbekannter Aufsatz Weberns über Schönberg', *ÖMz*, xxvii (1972), 127

V. Lampert: 'Schoenbergs, Bergs und Adornos Briefe an Sándor (Alexander) Jemnitz', *SM*, xv (1973), 355

A. Ringer: 'Schoenbergiana in Jerusalem', *MQ*, lix (1973), 1

F. Glück: 'Briefe von Arnold Schönberg an Claire Loos', *ÖMz*, xxix (1974), 203

K. Hicken: 'Schoenberg's "Atonality": Fused Bitonality?', *Tempo* (1974), no.109, p.28

A. Lessem: 'Schönberg and the Crisis of Expressionism', *ML*, lv (1974), 427

J. Maegaard: 'Schönberg hat Adorno nie leiden können', *Melos*, xli (1974), 262

J. Meggett and R. Moritz: 'The Schoenberg Legacy', *Notes*, xxxi (1974), 30

J. Samson: 'Schoenberg's "Atonal" Music', *Tempo* (1974), no.109, p.16

E. Schmid: 'Ein Jahr bei Schönberg in Berlin', *Melos*, xli (1974), 190

F. Schneider: 'Arnold Schönberg, Versuch einer musikgeschichtlichen Positionsbestimmung', *BMw*, xvi (1974), 75, 277

E. Steiner: 'Schoenberg's Quest: Newly Discovered Works from his Early Years', *MQ*, lx (1974), 401

R. Stephan: 'Hába und Schönberg', *Festschrift für Arno Volk* (Cologne, 1974)

W. Szmolyan: 'Schönberg in Mödling', *ÖMz*, xxix (1974), 189

J. Maegaard: 'Zu Th. W. Adornos Rolle im Mann/Schönberg-Streit', *Thomas Mann Gedenkschrift*, Text und Kontext, *Sonderreihe*, ii (Copenhagen, 1975)

A. Dümling: ' "Im Zeichen der Erkenntnis der socialen Verhältnisse": der junge Schönberg und die Arbeitersängerbewegung', *Zeitschrift für Musiktheorie*, vi (1975), 11

J. Harvey: 'Schönberg: Man or Woman?', *ML*, lvi (1975), 371

J. Maegaard: 'Der geistige Einflussbereich von Schönberg und Zemlinsky in Wien um 1900', *Studien zur Wertungsforschung*, vii (1976)

C. Dahlhaus: 'Schönbergs musikalische Poetik', *AMw*, xxxiii (1976), 81

E. Hilmar: 'Arnold Schönbergs Briefe an den Akademischen Verband für Literatur und Musik in Wien', *ÖMz*, xxxi (1976), 273

N. Nono-Schoenberg: 'Mon père Schoenberg', *SMz*, cxvi (1976), 2

W. Szmolyan: 'Schönbergs Wiener Skandalkonzert', *ÖMz*, xxxi (1976), 293

E. Steiner: 'Ein Schönberg-Konzert in Berlin', *ÖMz*, xxxi (1976), 105

J. Maegaard: 'Schönbergs Zwölftonreihen', *MF*, xxix (1976), 385

79

G. Mayer: 'Arnold Schönberg im Urteil Hanns Eislers', *BMw*, xviii (1976), 195

J. Rufer: 'Schoenberg – Yesterday, Today and Tomorrow', *PNM*, xvi (1977), 125

B. Simms: 'New Documents in the Schoenberg-Schenker Polemic', *PNM*, xvi (1977), 110

J. Theurich: 'Briefwechsel zwischen Arnold Schönberg und Ferruccio Busoni 1903–1915 (1927)', *BMw*, xix (1977), 163

A. Ashforth: 'Linear and Textural Aspects of Schoenberg's Cadences', *PNM*, xvi (1978), 195

A. Forte: 'Schoenberg's Creative Evolution: the Path to Atonality', *MQ*, lix (1978), 133

F. Schneider: 'Schönberg und die "politische Musik"', *BMw*, xx (1978), 23

W. Szmolyan: 'Schönberg und Eisler', *ÖMz*, xxxiii (1978), 439

C. Cross: 'Three Levels of "Idea" in Schoenberg's Thought and Writings', *CMc*, no.30 (1980), p.24

M. Hyde: 'The Roots of Form in Schoenberg's Sketches', *JMT*, xxiv (1980), 1

——: 'The Telltale Sketches: Harmonic Structure in Schoenberg's Twelve-Tone Method', *MQ*, lxi (1980), 560

A. Ringer: 'Weill, Schönberg und die Zeitoper', *Mf*, xxxiii (1980), 465

J. Smith: 'Schoenberg's Way', *PNM*, xviii (1980), 258 [recollections of people personally involved with Arnold Schoenberg and his circle]

P. Stadlen: 'Schoenbergs Speech-song', *ML*, lxii (1981), 1

W. Szmolyan: 'Die Konzerte des Wiener Schönberg-Vereins', *ÖMz*, xxxvi (1981), 82, 154

STUDIES OF PARTICULAR WORKS
(*operas*)

H. Keller: 'Schoenberg's "Moses and Aron"', *The Score* (1957), no.21, p.30

——: 'Schoenberg's Comic Opera', *The Score* (1958), no.23, p.27

K. H. Wörner: *Gotteswort und Magie* (Heidelberg, 1959; Eng. trans., rev., 1963, as *Schoenberg's 'Moses and Aron'*)

T. Wiesengrund-Adorno: 'Sakrales Fragment', *Quasi una fantasia* (Frankfurt, 1963), 306

K. H. Wörner: '"Die glückliche Hand", Arnold Schönbergs Drama mit Musik', *SMz*, cciv (1964), 274

H. Buchanan: 'A Key to Schoenberg's "Erwartung"', *JAMS*, xx (1967), 434

K. H. Wörner: 'Schönberg's "Erwartung" und das Ariadne-Thema', *Die Musik in der Geistesgeschichte* (Bonn, 1970), 91

Bibliography

J. Crawford: 'Die glückliche Hand: Schoenberg's Gesamtkunstwerk', *MQ*, lx (1974), 583

H. Boventer, ed.: *'Moses und Aron': zur Oper Arnold Schönbergs* (Bensberg, 1979)

E. Budde: 'Arnold Schönbergs Monodrama "Erwartung": Versuch einer Analyse der ersten Szene', *AMw*, xxxvi (1979), 1

S. Mauser: 'Forschungsbericht zu Schönbergs "Erwartung"', *ÖMz*, xxxv (1980), 215

A. Serravezza: 'Critica e ideologia nel "Moses und Aron"', *RIM*, xv (1980), 204

E. Steiner: 'The "Happy" Hand: Genesis and Interpretation of Schoenberg's Monumentalkunstwerk', *MR*, xli (1980), 207

R. Weaver: 'The Conflict of Religion and Aesthetics in Schoenberg's "Moses and Aaron"', *Essays on the Music of J. S. Bach and other divers subjects: a Tribute to Gerhard Herz* (Louisville, 1981), 291

(*choral works*)

A. Berg: *Arnold Schönberg: Gurrelieder Führer* (grosse Ausgabe, Vienna, 1913; kleine Ausgabe, 1914)

G. Strecke: 'Arnold Schönbergs op.XIII', *Melos*, i (1920), 231

S. Günther: 'Das trochäische Prinzip in Arnold Schönbergs op. 13', *ZMw*, vi (1923), 158

H. Nachod: 'The Very First Performance of Schoenberg's "Gurrelieder"', *Music Survey*, iii/3 (1950), 38

W. Zillig: 'Notes on Arnold Schoenberg's Unfinished Oratorio "Die Jakobsleiter"', *The Score* (1959), no.25, p.7

H. Pauli: 'Zu Schönbergs "Jakobsleiter"', *SMz*, cii (1962), 350

R. Lück: 'Arnold Schönberg und das deutsche Volkslied', *NZM*, Jg. 124 (1963), 86

K. H. Wörner: 'Schönbergs Oratorium "Die Jakobsleiter": Musik zwischen Theologie und Weltanschauung', *SMz*, cv (1965), 250, 333

C. M. Schmidt: 'Schönbergs Kantate "Ein Überlebender aus Warschau"', *AMw*, xxxiii (1976), 174

(*orchestral works*)

A. Berg: *Arnold Schönberg: Pelleas und Melisande Op. 5: kurze thematische Analyse* (Vienna, 1920)

C. Dahlhaus: *Schönberg: Variationen für Orchester, op.31* (Munich, 1968)

E. Doflein: 'Schönbergs Opus 16 Nr. 3', *Melos*, xxxvi (1969), 203

P. Förtig: 'Analyse des Opus 16 Nr. 3', *Melos*, xxxvi (1969), 206

J. Rufer: 'Nocheinmal Schönbergs Opus 16', *Melos*, xxxvi (1969), 366

C. Dahlhaus: 'Das obligate Rezitativ', *Melos/NZM*, i (1975), 193

P. Gülke: 'Über Schönbergs Brahms Bearbeitung', *BMw*, xvii (1975), 5

(*chamber works*)

[A. Berg]: 'Arnold Schönberg fis-moll-Quartett: eine technische Analyse', *Erdgeist*, iv (1909), 225

A. Berg: *Arnold Schönberg: Kammersymphonie Op. 9: thematische Analyse* (Vienna, 1918)

F. Greissle: 'Die formalen Grundlagen des Bläserquintetts von Arnold Schönberg', *Musikblätter des Anbruch*, vii (1925), 63

T. Wiesengrund-Adorno: 'Schönbergs Bläserquintett', *Pult und Taktstock*, v (1928), May–June, 45; repr. in T. Wiesengrund-Adorno: *Moments musicaux* (Frankfurt, 1964), 161

E. Schmid: 'Studie über Schönbergs Streichquartette', *SMz*, lxxiv (1934), 1, 84, 155

P. Gradenwitz: 'The Idiom and Development in Schoenberg's Quartets', *ML*, xxvi (1945), 123

W. Hyamson: 'Schoenberg's String Trio', *MR*, xi (1950), 184

O. Neighbour: 'Dodecaphony in Schoenberg's String Trio', *Music Survey*, iv (1952), 489

——: 'A Talk on Schoenberg for Composers' Concourse', *The Score* (1956), no.16, p.19 [on op.37]

W. Pfannkuch: 'Zu Thematik und Form in Schönbergs Streichsextett', *Festschrift Friedrich Blume* (Kassel, 1963), 258

E. Klemm: 'Zur Theorie der Reihenstruktur und Reihendisposition in Schönbergs 4. Streichquartett', *BMw*, viii (1966), 27

J. Lester: 'Pitch Structure Articulation in the Variations of Schoenberg's Serenade', *PNM*, vi/2 (1968), 22

E. Staempfli: 'Das Streichtrio Opus 45 von Arnold Schönberg', *Melos*, xxxvii (1970), 35

M. Pfisterer: 'Zur Frage der Satztechnik in den atonalen Werken von Arnold Schönberg', *Zeitschrift für Musiktheorie*, ii/1 (1971), 4 [on op.10]

U. von Rauchhaupt, ed.: *Schoenberg, Berg, Webern: the String Quartets, a Documentary Study* (Hamburg, 1971)

R. Gerlach: 'War Schönberg von Dvořák beeinflusst?', *NZM*, Jg.133 (1972), 122 [on the D major Quartet]

A. Whittall: *Schoenberg Chamber Music* (London, 1972)

——: 'Schoenberg and the "True Tradition": Theme and Form in the String Trio', *MT*, cxv (1974), 739

C. Raab: 'Fantasia quasi una Sonata: Zu Schönbergs "Phantasy for Violin with Piano Accompaniment" op.47', *Melos/NZM*, ii (1976), 191

Bibliography

C. Möllers: *Reihentechnik und musikalische Gestalt bei Arnold Schönberg: eine Untersuchung zum III.Streichquartett op.30* (Wiesbaden, 1977)

M. Hyde: *Schoenberg's Twelve-tone Harmony: the Suite op.29 and the Compositional Sketches* (Ann Arbor, 1982)

(solo vocal works)

R. Tenschert: 'Eine Passacaglia von Schönberg', *Die Musik*, xvii (1925), 590 [on op.21 no.8]

K. H. Ehrenforth: *Ausdruck und Form: Schönbergs Durchbruch zur Atonalität* (Bonn, 1963) [on op.15]

——: 'Schönberg und Webern: das XIV. Lied aus Schönbergs Georgelieder op.15', *NZM*, Jg.126 (1965), 102

C. Dahlhaus: 'Schönberg's Lied "Streng ist uns das Glück und spröde"', *Neue Wege der musikalischen Analyse* (Berlin, 1967), 45

H. Kaufmann: 'Struktur in Schönbergs Georgeliedern', *Neue Wege der musikalischen Analyse* (Berlin, 1967), 53

W. Stroh: 'Schoenberg's Use of the Text: the Text as a Musical Control in the 14th Georgelied, Op. 15', *PNM*, vi/2 (1968), 35

R. Brinkmann: 'Schoenberg und George: Interpretation eines Liedes', *AMw*, xxvi (1969), 1 [on op.15, no.14]

H. Weber: 'Schoenbergs und Zemlinskys Vertonung der Ballade "Jane Gray" von Heinrich Ammann: Untersuchungen zum Spätstadium der Tonalität', *IMSCR, xi Copenhagen 1972*, 705

A. Lessem: 'Text and Music in Schoenberg's "Pierrot Lunaire"', *CMc*, no.19 (1975), p.103

K. Bailey: 'Formal Organization and Structural Imagery in Schoenberg's "Pierrot Lunaire"', *SMA*, ii (1977), 93

H. Martin: 'A Structural model for Schoenberg's "Der verlorene Haufen" op.12/2', *In Theory only*, iii/3 (1977), 4

(keyboard works)

L. Welker: 'Arnold Schönbergs Op. 11', *Die Musik*, xii/1 (1912), 109

H. Leichtentritt: 'Arnold Schönberg: opus 11 and opus 19', *Musical Form* (Cambridge, Mass., 1951), 425

T. Tuttle: 'Schoenberg's Compositions for Piano Solo', *MR*, xviii (1957), 300

A. Forte: 'Context and Continuity in an Atonal Work', *PNM*, i/2 (1963), 72 [on op.19]

W. Rogge: *Das Klavierwerke Arnold Schönbergs* (Regensburg, 1964)

R. Travis: 'Directed Motion in Schoenberg and Webern', *PNM*, iv/2 (1966), 85 [on op.19/2]

R. Wille: 'Reihentechnik in Schönbergs opus 19, 2', *Mf*, xix (1966), 42

G. Krieger: *Schönbergs Werke für Klavier* (Göttingen, 1968)
R. Brinkmann: *Arnold Schönberg: Drei Klavierstücke Op. 11* (Wiesbaden, 1969)
J. Graziano: 'Serial Procedures in Schoenberg's Opus 23', *CMc*, no.13 (1972), p.58
H. Oesch: 'Schönberg im Vorfeld der Dodekaphonie', *Melos*, xli (1974), 330 [on op.23 no.3]
K. Bailey: 'Row Anomalies in op.33: an Insight into Schoenberg's Understanding of the Serial Procedure', *CMc*, no.22 (1976), p.42
J. Glofcheskie: ' "Wrong" Notes in Schoenberg's op.33a', *SMA*, i (1976), 88
D. Gostomsky: 'Tonalität-Atonalität: zur Harmonik von Schönbergs Klavierstück op.11 Nr.1', *Zeitschrift für Musiktheorie*, vii/1 (1976), 54
J. Maegaard: 'Om den kronologiske placering af Schönbergs klaverstykke op.23 nr.3', *Musik en forskning*, ii (1976)
D. Stein: 'Schoenberg's op.19 no.2: Voice leading and Overall Structure in an Atonal Work', *In Theory only*, ii/7 (1976), 27
W. Grangjean: 'Form in Schönbergs op.19, 2', *Zeitschrift für Musiktheorie*, viii/1 (1977), 15
M. Guck: 'Comment: Symmetrical Structures in op.19, 2', *In Theory only*, ii/10 (1977), 29
D. Lewin: 'Some Notes on Schoenberg's op.11', *In Theory only*, iii/1 (1977), 3
T. Bond: 'Schoenberg's Sonata for Organ', *MT*, cxix (1978), 984

(theoretical works, etc)
E. Stein: *Praktischer Leitfaden zu Schönbergs Harmonielehre* (Vienna, 1923)
R. Lück: 'Die Generalbass-Aussetzungen Arnold Schönbergs', *DJbM*, viii (1963), 26
C. Parmentola: 'La "Harmonielehre" di Schoenberg nella crisi del pensiero moderno', *NRMI*, ii (1968), 81
L. Richter: 'Schönbergs Harmonielehre und die freie Atonalität', *DJbM*, xiii (1968), 43
D. Rexroth: *Arnold Schönberg als Theoretiker der tonalen Harmonik* (Bonn, 1971)
J. Spratt: 'The Speculative Content of Schoenberg's "Harmonielehre" ', *CMc*, no.11 (1971), 83
R. Stephan: Schönbergs Entwurf über "Das Komponieren mit selbständigen Stimmen" ', *AMw*, xxix (1972), 239
C. Dahlhaus: 'Schoenberg and Schenker', *PRMA*, c (1973–4), 209
A. Goehr: 'The Theoretical Writings of Arnold Schoenberg', *PRMA*, c (1973–4), 85

Bibliography

(*painting*)

J. Rufer: 'Schönberg als Maler, Grenzen und Konvergenzen der Künste', *Aspekte der neuen Musik*, ed. W. Burde (Kassel, 1968)

W. Hofmann: 'Beziehungen zwischen Malerei und Musik', *Schönberg, Webern, Berg: Bilder, Partituren, Dokumente* (Vienna, 1969)

WEBERN

Paul Griffiths

CHAPTER ONE

Life

Anton Friedrich Wilhelm von Webern was born in Vienna on 3 December 1883. His childhood was spent in Vienna, Graz (1890–94) and Klagenfurt, where he had piano and cello lessons with Edwin Komauer. He composed some of his first pieces at the family's summer estate of Preglhof in Carinthia; and there he and his cousin Ernst Diez passed the time in walking and in collecting plants and minerals, mountain pursuits for which Webern retained a fondness throughout his life. In 1902, on graduating from the Klagenfurt Gymnasium, he was rewarded by his father with a trip to Bayreuth. That autumn he entered the University of Vienna to study musicology under Adler, also taking harmony with Graedener and counterpoint with Navratil, and continuing his studies of the cello and piano. At the same time he sang in the Akademischer Wagner Verein under Mottl, Richter, Nikisch and Mahler. In the autumn of 1904 he began lessons with Schoenberg.

For Webern, as for many others, Schoenberg's teaching was of decisive importance. His compositions of 1904–5 demonstrate how much he owed to the purely technical instruction he received from Schoenberg, but just as significant was the contact with Schoenberg's philosophy and the mutual friendship and respect that developed between teacher and pupil; their relations remained close until Schoenberg's departure for Berlin in 1925. Webern also struck up warm and lasting friendships with his co-pupils Wellesz and Berg, whose death in 1935 affected him profoundly. In 1906 he took a DPhil for his work on Isaac, and his edition of the second part of the *Choralis constantinus* was published three years later. Lessons with Schoenberg continued until 1908, when he presented an informal graduation exercise: the Passacaglia op.1 for orchestra.

Webern now began a rather precarious conducting career, for which he had had no training. He held appointments at Bad Ischl (summer 1908), Teplitz (1910) and Danzig (1910–11), conducting principally operettas and other light music. In autumn 1911 he followed Schoenberg to Berlin; he attended the first performance of *Das Lied von der Erde* in Munich in November; and in the following June he moved to a conducting post in Stettin, which he abandoned early in 1913. After experiencing various military training courses (1915–17) he took a conducting engagement at the Deutsches Theater in Prague (autumn 1917), but again his tenure was short: in August 1918 he returned to Vienna and settled close to Schoenberg in Mödling. He took a leading part in the Verein für Musikalische Privataufführungen, playing the cello, directing performances and making arrangements of his own and other pieces. In 1921 his scores began to appear from Universal Edition.

It was at this time that Webern's serious conducting activity began: he assumed the direction of the Schubertbund (1921–2), the Mödling Male Chorus (1921–6), the Vienna Workers' Symphony Concerts (1922–34) and the Vienna Workers' Chorus (1923–34). In his work with the last two associations, in particular, he was stimulated by an idealistic wish to

8. *Black
chalk drawing
(1918) of
Webern by
Egon Schiele*

encourage popular education and the appreciation of
new music, though of course he also conducted the
Viennese classics from Haydn to Brahms. Apart from
his own works and arrangements, his programmes
included several Mahler symphonies and pieces by Berg,
Reger and Schoenberg; on one occasion he conducted
an American concert, including music by Ives. In 1926
he was invited to the ISCM Festival to direct performances of Schoenberg's op.26 and his own op.10; in

91

1927 he was given a regular conducting appointment with Austrian radio; and in 1929 he made his first important tour, visiting Munich, Frankfurt, Cologne and London. His letters and the reports of contemporaries indicate that as a conductor he was meticulous in his approach to the score.

Other events of these years included two awards of the Viennese Grosser Musikpreis (1924, 1932) and Webern's meeting with the poet Hildegard Jone (1926), who provided the texts for all of his completed vocal works after this date. Also in 1926 he accepted a post as teacher of music theory at the Jewish Cultural Institute for the Blind. This was his only formal teaching appointment, though he did have a number of private pupils, among them Hartmann, Searle and Spinner. For his pupils he gave two series of lectures: 'Der Weg zur Komposition in zwölf Tonen' (1932) and 'Der Weg zur neuen Musik' (1933), both published under the latter title in 1960.

The advent of Nazism brought the dissolution of the workers' associations in 1934 and the withdrawal of Webern's radio post in 1938, so that he was forced to take routine work from Universal Edition. He spent most of the war in isolation at Mödling, though in 1943 he was able to attend the première in Winterthur of his Variations op.30 for orchestra. In the last weeks of the war he and his wife left Vienna to be with their daughters at Mittersill, in the mountains near Salzburg. He died there on 15 September 1945, having been shot in error by an American soldier.

CHAPTER TWO

Towards atonal concision 1899–1914

Most of Webern's 'pre-Schoenberg' compositions are hardly more than juvenile exercises (many have been published in performing editions), showing how much and how rapidly he was to gain in accomplishment under Schoenberg's guidance during the years 1904–8. The String Quartet of 1905 already shows a striking advance. Webern's model here was *Verklärte Nacht*, which is almost quoted at some points. Like Schoenberg he tried to weld sections of quite different speeds, textures and tonalities into a single movement, but the joins are too many and awkward for this to be achieved successfully, and the varied recurrences of a three-note motif are not sufficient to strengthen the structure.

With the Piano Quintet (1907) Webern produced a work firmer in form and tonality: it can be regarded as a sonata movement in C major and in a somewhat solid Brahmsian style. Brahms again, now seen from the standpoint of Reger and Schoenberg, lies behind the Passacaglia op.1 for orchestra (1908). This is a marvellously continuous essay in Schoenberg's 'developing variation', a technique to which the form is well suited, though the passacaglia theme is lost as the music surges to its climax. In orchestration the piece draws close to Mahler and to the Schoenberg of *Pelleas*, particularly at

93

its moments of churning rhetoric; but such passages as the first variation – scored *pianissimo* for flute, trumpet, harp, violas and cellos – are quite distinctively Webernian.

Later Webern recalled the awed shock with which he had received Schoenberg's First Chamber Symphony (1906). In the Passacaglia he held back from the harmonic adventurousness of that work, keeping about five years behind his teacher as he had in the String Quartet of 1905. But at the same time he was completing a set of five Dehmel songs (1906–8) which show him abreast of Schoenberg's tentatives towards atonality. Each of the songs has a key signature, but the tonality is generally dissolved out of significance, often by the use of diminished triads and other whole-tone formations. The vocal line is quite undeclamatory, moving hesitantly in small steps within a restricted compass, and there is a good deal of imitation in the accompaniment. In lightness of harmony and texture these songs are at one with the later Webern, bringing, for example, an appropriate icy clarity to Dehmel's vision of the soul as a star drinking eternal light in 'Am Ufer'. The most impressive number is 'Helle Nacht', a song in triple counterpoint with the parts varied and differently assigned (between voice, left hand and right hand) in each of the three sections.

Like Schoenberg, Webern began to explore 'the air of another planet' with the support of words by George, which he used in the chorus *Entflieht auf leichten Kähnen* op.2 (1908) and in 14 songs of 1907–9 from which he published ten as opp.3 and 4. George's invocations to leave behind the world of appearances for that of dream and fantasy (op.2, op.3 no.5, op.4 no.1) were an obvious parallel for new musical departures, just as his dislocated syntax fitted well with music that was all suspension. The choral piece, a four-part canon, still has a key signature, but the tonality drifts almost from chord to chord. With the songs of opp.3 and 4 Webern abandoned reference to tonality or, for the most part, rhythmic stability: the music is pulseless, flexible in tempo and irregular in phrase length, metre and rhythmic pattern. The dynamic level rarely rises above *piano*, the pieces are mostly very short and close repetition is avoided. All of these features were to remain characteristic of Webern's early atonal music. In the case of opp.3 and 4 they lend an intense lyrical intimacy to the settings: 'Dies ist ein Lied für dich allein', the first line of op.3 no.1, might be a motto for all ten. But the intimacy is of loss as much as love; the poems are filled with images of cold, greyness and winter, echoed in the homeless (keyless) music and its stripped textures.

As has been implied, these songs of opp.3 and 4 are freely formed (Webern made little attempt to shape his songs on George's superbly finished verbal structures), and their moment-to-moment coherence is similarly hard to grasp; imitation, for example, is more subtle than in the Dehmel songs, and the contrapuntal impetus of that earlier set is generally diminished, particularly in op.4. However, it is possible to detect the importance of basic motifs (properly, pitch-class sets) which persistently reappear transformed by inversion, transposition, reordering, octave displacement and/or presentation (linear or chordal). Again, this is an important technique in Webern's later atonal music.

Abandoning tonality, Webern and his colleagues

found it necessary to abandon thematic working as well, since their understanding of development was linked inextricably with tonal modulation. Webern now faced the problem of creating purely instrumental atonal athematic music, composing in 1909–10 the Five Movements op.5 for string quartet, the Six Pieces op.6 for orchestra and the Four Pieces op.7 for violin and piano. The only one of these pieces in which thematic development has a major role is op.5 no.1, a sonata movement treating a few motivic ideas; and at 55 bars it is the most extended number. In the others the form is more elusive, even where it is shaped by a progressive accumulation (op.6 nos.2 and 4) or an increase then decrease in contrapuntal density (op.6 no.1). Several of the pieces are so short that there is no room for anything but the statement of some brief ideas bearing a shadowy relationship to one another.

Each set is balanced between pieces of two sorts: those that are consistently restrained in dynamic, melodic range, colour and harmonic type; and those in which some or all of these and other aspects are abruptly variable. In op.7 no.4 the two are brought together, so that a few bars of calm stillness abate the passionate tension that has been swiftly established. These bars contain an element that is characteristic of the subdued music of opp.5–7: a phrase of about six notes that ascends or descends in regular rhythm, with intervals of a 7th or a minor 9th divided by one or two notes (in this descending example, marked 'as a breath', the pitches are D–Bb–Eb–B–G–C–F♯). Semitones and their octave transpositions are preponderant in the melody and harmony throughout all 15 pieces, and small semitone-containing sets can often be detected as fundamental structural elements.

Whether in frenzy or delicate quiescence, Webern's timbral discrimination is extremely important. Op.6 is scored for an orchestra of similar constitution to that of Mahler's Sixth Symphony, but these forces are used to weigh out massive sonorities only in the second and fourth pieces, the latter being an awesome funeral march for wind and percussion. In the remaining numbers, varied groupings are employed and the colouring is much less consistent. This instability of timbre is particularly marked at the opening of the first piece, with the melodic line passing between instruments (flute–horn–flute–trumpet). Perpetual change is indeed almost a principle at all surface levels in opp.5–7. Sometimes there is a constant metre, most notably in op.6 no.4 and op.5 no.3, but more often this is lacking, either because the prevailing time signature is obscured (op.6 no.1) or because it shifts frequently (op.6 no.3 and op.7 no.3). With such pieces as these last two, of ten and 13 bars respectively, Webern might have appeared to have been at the end of the road: there is no development, only obsessive repetition or a fleeting passage from motif to motif.

At this point Schoenberg found support for larger forms in dramatic texts (*Erwartung* and *Die glückliche Hand*). Webern had briefly planned an opera in the very different circumstances of 1908, a setting of Maeterlinck's *Alladine et Pallomides*, of which a single-page sketch survives; but now he turned to brief, intimate poems by Rilke, mirroring the state of momentary essence that he had reached in the most fugitive pieces of opp.5–7. The voice in these Two Songs op.8 (1910) provides a thread of continuity, yet the accompaniment, for eight instruments, is still more fragmented and fast-moving than anything in the immediately preceding compositions, touching the melody with tiny dashes of music. The melody itself has all the graceful ease of opp.3 and 4 while wandering through a wider range and in wider intervals: the 7ths and minor 9ths that were to remain characteristic of Webern's lines are more exposed than before.

The next three works parallel the triad of 1909–10: again there is a set of compositions for string quartet – the Six Bagatelles op.9; one for orchestra – the Five Pieces op.10; and one for string instrument and piano – the Three Little Pieces op.11. But the concision of these pieces is such as to make their predecessors appear colossal: a few motifs, chords and ostinatos, and the music is past. Something of the anguish that caused and was caused by this brevity was recalled by Webern in speaking of the Bagatelles: 'While working on them I had the feeling that once the 12 notes had run out, the piece was finished. . . . It sounds grotesque, incomprehensible, and it was immensely difficult'.

Almost the closest he came to a literally 12-note composition was op.10 no.4, which is so short and thinly scored that it may be quoted complete (ex.1). Here the chromatic total is presented in the first 12 notes, but the music is not serially composed. It is possible to see the importance of pitch-class sets (the first three notes in the mandolin form one which reappears several times, and at significant junctures), as in all of Webern's non-serial atonal compositions; and yet the

Ex.1 Five Pieces for Orchestra, op. 10 no. 4

structure of this music remains enigmatic. Perhaps the
only obvious pattern is that formed by the last intervals
of each of the four melodic fragments.

Though more than usually brief, op.10 no.4 is characteristic of this group of three works and shows the
culmination of tendencies in Webern's music since op.3.
Linear polyphony is hardly present at all, the music is
virtually ametric, the texture is open, the timbre changes
from bar to bar, the dynamics are attenuated, and
phrases are of no more than six notes. The public

rhetoric of op.6 has been left, along with its enormous forces: op.10 is scored for no more than 20 players, with an approximate balance among wind, strings and percussion. There is something of late Mahlerian orientalism in the complement, but not in Webern's highly fluctuating use of it. The first three notes of op.10 no.1, each of which is differently orchestrated, show this refinement at its peak (in a similar way op.6 no.1 had begun with an extreme of timbral discontinuity); though such precedents for what was later termed 'pointillism' are few, and the scoring of op.10 no.4 is much more typical.

Webern's plan for the set eventually published as op.10 changed several times: three of the posthumously published Four Pieces relate to this work (the other is close to op.6), and one scheme included the orchestral song 'O sanftes Glühn der Berge' of 1913. This has a text by Webern himself concerning a vision of a woman; and in the same year he wrote another song to his own words, 'Schmerz, immer blick' nach oben', for voice and string quartet (associated with the Bagatelles), and also a drama, *Tot*. The last was prompted immediately by the death of a nephew, but Webern was still feeling the loss of his mother, who had died in 1906. As he wrote to Berg in 1912: 'Except for the violin pieces and a few of my orchestra pieces, all of my compositions from the Passacaglia on relate to the death of my mother'. If regret and abandonment are the predominant expressive qualities of opp.1–11, it was Webern's achievement to have created music that could be at once mournful and extraordinarily intense.

The possibility of that combination is not unrelated to Webern's compression, which reached the furthest point

9. *Autograph MS of 'Three Little Pieces', op.11 no.3, composed 1914*

in the Three Little Pieces for cello and piano (1914): the last consists of just 20 notes. Along that road lay, indeed, no more than a permutation of the chromatic scale. The composition of the cello pieces was immediately preceded by that of the first movement of a sonata for the same instruments, a work in which, as Webern wrote to Schoenberg, he had attempted 'to find at last an approach to longer movements'. This movement (a projected second was not written) is still quite brief, but it is more developed than anything in

101

Webern's music since op.5 no.1. However, the difficulties in writing atonal athematic music without the formal substrate of a text were unresolved, and apart from a movement for clarinet, trumpet and violin begun and abandoned in 1920, Webern composed no more purely instrumental music until 1924 – immediately after the introduction of serial technique.

CHAPTER THREE
Towards serialism 1914–27

During World War I Webern added to 'O sanftes
Glühn' several more songs accompanied by small
orchestral forces (like op.10), but the four he chose for
publication as op.13 did not include those to his own
texts. Concurrently he composed four songs with piano
accompaniment, published as op.12. After an extremely
rapid advance – within three years of the Passacaglia he
had been at work on op.10 – this was a time of steadier
development and less prolific creativity. The vocal style
takes up from op.8 with little change, but the songs of
opp.12 and 13 are longer and, from the earliest
onwards, their accompaniments become more linear.

In style and subject the op.12 set makes a heterogeneous group. The folksong setting 'Der Tag ist vergangen' is light and simple, the Goethe song 'Gleich und
gleich' almost playful; but in a sense these are the most
forward-looking, since their themes – naive piety and
the observation of natural principles – were to dominate
Webern's later vocal music. 'Schien mir's, als ich sah die
Sonne', much the most declamatory song of the period,
sets a mystic, moral poem from the *Ghost Sonata* of
Strindberg, who was one of the writers most important
to Webern, though this was his only setting. 'Die
geheimnisvolle Flöte' has a delicate sultry exoticism, a
manner Webern also used for two other settings of

Bethge's translations from the Chinese, the central numbers of op.13. 'Die Einsame' (1914), the earliest of this set, bears comparison with Webern's contemporary settings of his own verse, 'O sanftes Glühn' and 'Leise Düfte', in its instrumental fragmentation, while the later numbers are more polyphonic. However, op.13 draws unity from its concern with the figure of the wanderer, exiled from a distant home, his longing intensified by all he sees of nature (represented in the Chinese poems by the moon, though parallels with Schoenberg are scarcely tenable). In Trakl's 'Ein Winterabend', the last song in time (1918) and placing, the traveller finds rest in an appearance of the sacrament.

Webern had already set another Trakl poem, 'Abendland III', in a quite different style: thoroughly contrapuntal and a great deal more forceful in expression. Five more settings were added to form the Six Songs op.14 for soprano, two clarinets and two strings (1917–21). Throughout most of the set the instrumental lines are as continuous as the voice part, which takes a more tortured course than any in Webern's previous output: the range is wide and the dynamic requirements are particularly demanding. The instrumentation is different in each number, but it is consistently dark and brooding. In all of this the influence of *Pierrot lunaire* has been discerned, but there is nothing theatrical about these pieces. Indeed, Webern does nothing to accentuate the horrific imagery of decay in the texts: it is the abandoned melancholy of Trakl's verse that is emphasized, not his vision of the immanence of evil.

The new features of the Trakl songs – a soprano part frozen in tension, accompanied by packed instrumental polyphony – were pursued in the first four of the Five Sacred Songs op.15, composed in 1921–2. As for op.12 no.1, Webern chose devotional folk poems, but the treatment is here very far from naive. The last of the set, a much more serene piece written in 1917, is a double canon, and so looks forward to Webern's next work, the Five Canons on Latin Texts op.16 for soprano, clarinet and bass clarinet. In opp.14 and 15 counterpoint had flowed with a richness and freedom unparalleled in Webern's music; he now resorted to structures of bare strictness. In other ways, too, the music is made more firm and stripped: lines are composed mainly of short phrases separated by rests, rhythms are more regular than hitherto, the flexibility of tempo and metre is reduced, and there are sharp contrasts in dynamics. Only the second piece is calm and quiet throughout, in accord with its simple text from *Des Knaben Wunderhorn*. The others, setting extracts from the Holy Week liturgy, bear the extreme in jaggedness of Webern's vocal writing. Polyphonic textures were henceforth to be the rule, but Webern was not to repeat the penitential starkness of these canons.

1924, the year in which the Latin canons were completed, saw Webern's first successful return to instrumental composition with the tiny *Kinderstück* for piano, a simple exercise in serialism, with the series repeated always in the same form. The next year he used the same method on a somewhat larger scale in the Piano Piece and the Movement for string trio. Of all three, the last is the most noteworthy. It is very short, but its athematic handling of the series, which generates a diversity of motifs presented in rapid equal notes, is comparable with that in the first serial instrumental work Webern was to publish, the String Trio op.20 of 1926–7.

Meanwhile he was continuing the sequence of vocal cycles with accompaniments for small instrumental ensembles. The Three Traditional Rhymes op.17 (1924–5) and the Three Songs op.18 (1925) proceed directly from opp.14–16 in style: the vocal line is wide-ranging and highly contorted, reaching an extreme of tense utterance in op.18 no.2, where the voice leaps back and forth, often in 7ths and minor 9ths, over a span of two octaves and a 5th; and the instrumental parts are equally stretched to the limit, particularly that for E♭ clarinet in op.18. There is also a link with op.16 in the choice of texts. Those of op.17 are popular prayers for redemption, and the theme is taken up with renewed intensity in 'Erlösung' op.18 no.2, a setting of dialogues of entreaty between the Virgin and Christ and between Christ and the Father. The third song of op.18 sets the hymn 'Ave regina coelorum', but the first is, seemingly, a simple love-song. However, for Webern, as he wrote to Berg, the three poems were linked, and op.18 no.1 might be interpreted as an assurance of salvation as much as of earthly matrimony, making the group a cycle on the intercession of the Virgin.

It is possible to trace the early development of Webern's serial technique in the six numbers of opp.17 and 18: op.17 no.1 is non-serial; op.17 no.2 and op.18 no.1 (the first of the later set to be composed) are built, like the contemporary instrumental pieces, from abutting statements of the same serial form; op.17 no.3 repeatedly states the series in the voice, complementing it with serial fragments in the accompaniment; op.18

no.2 uses four forms – a prime and its inversion starting on the same note, together with the retrogrades of both – to distinguish the four separate lines of speech in the text; and op.18 no.3 is the first example of the technique that was to become standard for Webern, with each contrapuntal line following through its own sequence of serial forms. The impression is of an examination of different possibilities of the method, no two of the five serial songs being based on the same series. But the use of serialism by no means determines style. All the songs of 1922–5 treat spiritual themes in an acutely strained, fully chromatic manner, for which serial procedures perhaps provided the framework of rule that Webern had at first sought in canonic writing.

The style is more relaxed in Webern's next composition, the Two Songs op.19 (1926) on texts from Goethe's *Chinesisch-deutsche Jahres- und Tageszeiten*, set for chorus accompanied by violin, two clarinets, celesta and guitar. Here the instrumental writing is more than usually full of the rapid repeated notes characteristic of Webern's early serial works, giving the pieces an exquisitely fussy quality. In serial technique op.19 makes two advances: both numbers are based on the same series, and transpositions are used. A fascinating sketch survives for a third Goethe chorus, without accompaniment. The few bars that Webern set down show the beginnings of a four-part canon, each voice announcing its own serial form – a brief glimpse forward to the pure sobriety of the Symphony op.21.

CHAPTER FOUR
Serial works 1927–45

I Instrumental works 1927–40

Webern was now able to return to instrumental composition on a scale that he had not been able to sustain since the Passacaglia. As he was to recall: 'Only after Schoenberg pronounced the law did larger forms become possible again'. Even so, his first extended serial composition, the String Trio op.20 (1926–7), is typically compact, lasting for about eight minutes. Its two movements can be considered as a rondo and a sonata, but those forms are by no means immediately perceptible. The notes of the rondo theme, for instance, are reinterpreted in terms of register, timbre, rhythm and dynamic at each of the three repetitions; and the theme is a polyphonic one. A comparison with Schoenberg's contemporary Third Quartet shows how far the two were apart in their approaches to formal convention. In the case of Webern, there is little in the music that suggests an accompanied melodic theme, and the surface coherence depends less on thematic development than on consistency of rhythmic shape and interval. As in the songs of opp.17 and 18, there is a preponderance of semitone intervals, chiefly 7ths and minor 9ths, but also larger spans. And the tight-pressed, extreme activity of the music also relates to the immediately preceding works rather than to those that were to follow.

What distinguishes the instrumental compositions of 1928–40 from the String Trio is, in a word, symmetry. The Symphony op.21 already shows this in the forms of its two movements: symmetry about a horizontal axis in the four-part mirror canon of the first; symmetry about a vertical axis in the palindromic variations of the second. In the works that followed – the Quartet op.22 for violin, clarinet, tenor saxophone and piano, the Concerto op.24 for nine instruments, the Piano Variations op.27, the String Quartet op.28 and the Orchestral Variations op.30 – canons, palindromes and variations recur as principal structures. Often these formal types are combined, with each other or with different groundplans: sonata form, rondo and so on. The first movement of the String Quartet is typical, in that it is constructed as a theme and six variations which also functions as an 'adagio form' (theme = introduction, variation 1 = theme, variation 2 = transition to second subject, etc), and at the same time it is, except for the first 15 bars, composed in two-part canon.

As in the String Trio, Webern's notion of a theme is unconventional. Only at the opening of the Symphony's second movement and in the second section of the Orchestral Variations is there a clearly exposed melodic theme, and in these cases the identity of the theme is entirely lost in the subsequent development. More typically, a Webern subject consists of a collection of motifs (harmonic, melodic or rhythmic), often differentiated in timbre and/or rhythmic shape, and usually presented in a very open texture; the initial bars of opp.22, 24 and 30 show this most clearly. The subject itself does not return (except, of course, in marked repeats) without considerable change, even in a recapitulation such as that of the first movement (sonata) of op.22; but apprehension of Webern's themes as such makes his larger planning the more clear and satisfying.

That apprehension is assisted, as it is not in the String Trio, by the fact that a theme is usually built from closely related motifs. The theme of the op.22 movement mentioned – ten bars of two-part mirror canon threaded by a saxophone melody – shows how much is repeated within a subject, here built from clearly distinguished two- and three-note motifs. This omnipresence of the same simple ideas reaches an extreme point in the Concerto, where everything derives from one three-note motif, perpetually varied by transposition and/or inversion and/or retrogression, and ever presented afresh (instrumentally, harmonically/melodically and so on). Where the basic motif is as pervasive as it is here, the problem arises that its use to create larger units may not be perceived, the riveting coherence at the surface level blocking a wider understanding. For example, the play of two- and three-note groups in the second movement of op.24 can divert attention from the patterning of units of two or three bars. There is thus a subtle relationship between small and large structures, the former being necessary to, but also compromising, perception of the latter.

This is particularly evident in Webern's canons. That of the Symphony, for example, is not presented in four distinct voices: each part changes repeatedly in timbre and register, so that its uniqueness is less immediately apparent than are its shared properties, the short sequences it has in common with other lines. A dramatic instance occurs in bars 47–8, where a rising B–B♭ is heard twice in the first violin part, first solo *piano* and then tutti *forte*: this progression is far more striking than the provenance of the dyads from different canonic voices. The middle movement of the Piano Variations goes much further in destroying canonic appearances, though, perhaps paradoxically, the structure depends very much on the existence of a canonic relationship. Indeed, it would be curious if Webern had composed canons and then wilfully atomized them out of perceptibility.

In this respect a statement of his on the Bach transcription he made at this period (1934–5) is pertinent. Webern had orchestrated the six-part ricercare from *Das musikalische Opfer*, using the technique of motif separation that appears in his own Symphony and Orchestral Variations. In 1938 he wrote to Scherchen, who was to conduct the arrangement: 'the theme throughout must not appear disintegrated. My orchestration tries (here I am speaking of the whole work) merely to reveal the motivic coherence'. It may not be too presumptuous to apply this comment to his original compositions, where, again, the performer must be able to maintain the continuity of a line despite changes in register or timbre. Support for such an extension of the remark comes from Webern's reported reaction to an unsatisfactory performance of the Symphony: 'A high note, a low note, a note in the middle – like the music of a madman!'.

But, whatever the performer may do, a Webern theme is a more than usually diverse object, to the extent that its distinction from any development and its recognition in any recapitulation (without, as has been noted, exact restatement) becomes just as problematic as its identity.

Webern's solutions to the difficulties in articulating large-scale structures were several. Some are shown straightforwardly in the first movement of the Symphony. The opening part (up to bar 25) is slow and even (apart from grace notes there are only durations of one to four or six crotchets), and it is contained in a static medium-low register (there are only 13 pitches, ranging over three octaves and a tone up to *e''*). The second part introduces quavers and gradually increases the compass to *c♯''''*, undermining at the same time the harmonic stasis, though this last feature returns in the latter half. Thus the movement has qualities of exposition (stability), development (instability) and recapitulation (return to stability) without tonal harmony. The first movement of the Quartet op.22 uses register, and also density, to similar ends. (Something has been made of the 'symmetrical harmony' in these and other movements; but, if each pitch class is fixed in register and the composition is a canon by inversion, symmetrical harmony is bound to be produced. The important feature would seem to be the fixity of register. However, see Archibald (1972) for a claim that symmetrical harmony is audibly functional in op.5 no.2.)

Just as tempo, register and density are used to elucidate large form, so the motivic workings of these late instrumental pieces are made clearer by the use of colour, dynamics, duration and rhythm (there is little evidence to support an autonomous 'serial' organization of non-pitch elements). The primacy of pitch is indicated by the avoidance of means that produce indeterminacy in this area. In extreme contrast with the String Trio and the earlier quartet pieces, op.28 contains no harmonics, no *col legno*, and only a very little *sul ponticello* playing. Similarly, there are no unpitched percussion instruments in the orchestral compositions (the choral works opp.26, 29 and 31 are also highly cautious in the use of noise). The orchestration is, however, as light, fleeting and graceful as that of Webern's most characteristic earlier pieces, particularly in the Variations, where, in general, the writing is freer and more varied than in the other works of this group.

One of the happiest aspects of all these compositions is their playing with rhythm. Pulse is not often a strong force in Webern's atonal works, and the often changing metres of the String Trio produce little result, given the continual cross-rhythms and syncopations. The Symphony, however, does a great deal with oppositions between motivic rhythm and fixed metre, and the Quartet op.22 owes something of its playful character to skill in metric manipulation, used to considerable effect in all the subsequent works. Webern's success in some cases has, however, been called in question, even by such an authority as Stravinsky. As for the serial technique in these works, which have been the most extensively analysed from this point of view, further observations will be found in chapter 5.

II Songs and cantatas 1929–45

All the vocal works that Webern completed after his contact with Hildegard Jone (1926) are to texts by her. Collaboration on a stage work had been mooted by January 1930, but that plan was dropped in the following September when Webern asked for a cantata text. This too came to nothing and, in the event, he made his selections from existing poems for the six songs of 1933–5 and the three choral pieces of 1935–43.

During this period he was in regular correspondence with Jone and her husband, the sculptor Josef Humplik, discussing his work and theirs. It is clear from his letters that he was deeply impressed by Jone's verse, which, though of no great literary quality, gave him the verbal and philosophical materials he required: a view of nature as displaying in its order and symmetry the grace of God, imagery drawn from the lives of insects and plants, a vision of the human soul as a source of warmth and light, and faintly mystic Christian piety.

Webern had made sketches for three songs in 1929–30, but the earliest Jone settings, the second and third numbers of op.23, were the first vocal works he had completed for seven years. In the interim his open-textured, motivic contrapuntal style had been developed, and the two songs show something of this, though it is handled with more freedom than in the surrounding instrumental pieces. Certainly there is a great change from the anguish of opp.16–18: the vocal line returns to the simplicity of tone and beauty of melody – qualities that suggest a readier comparison with Schubert than with any of Webern's contemporaries – that distinguished some of the songs of opp.12 and 13; and the strenuous instrumental counterpoint of the middle-period songs is replaced by a light piano accompaniment. The op.23 set was completed in 1934 with 'Das dunkle Herz', one of the longest of Webern's songs, being formed as a binary aria. Within a year he had finished another three songs (op.25) in a similar style, beginning while he was at work on the Concerto, which at one stage, it appears from the sketches, was to have included a choral movement to a Jone text.

The choral project was realized in the next composition, *Das Augenlicht* op.26: Webern chose the poem with the plan for a work for chorus and orchestra already in mind. In 1938–9 it was followed by the First Cantata op.29, of which the central movement, a soprano aria, was composed before the two choral numbers. Both works, to some extent, relate stylistically to the contemporary instrumental pieces, although *Das Augenlicht* is freer in form than anything else of this period. The music is predominantly in four-part counterpoint (the chorus varies this with passages of homophony), the instrumental lines are distributed between instruments, and many other features noted in the previous section may be found. In the aria of op.29, the soprano sings one part of a canon, with the other three parts divided into a skein of fragments around the voice, rather in the manner of the first movement of the Symphony. And the last number of the cantata is another example of the formal fusion described in the String Quartet: Webern considered it at once a fugue, a scherzo and a set of variations. In vocal style, despite the frequent occurrences of wide intervals, opp.26 and 29 continue the pure tranquillity of the preceding songs.

Webern's last composition, the Second Cantata op.31 (1941–3), is his longest, with a duration of a little more than ten minutes, and requires larger forces than any work since op.6. The plan of the cantata came about during the course of composition: Webern began with the fourth movement and wrote the others in the order 5, 6, 1, 2, 3, choosing his texts from different collections. A seventh section was abandoned, and the scheme was altered at least once before the final version was arrived at: bass recitative, bass aria, female chorus with solo soprano, soprano recitative, soprano aria with chorus and obbligato violin, chorus. Webern likened the form to that of the Mass Ordinary, and this, together with the look of the score – all of the works from the Symphony onwards tend to use long note values in short bars, and the final chorus of this cantata has the four voices in different metres – has led to the drawing of parallels with Renaissance music. However, op.31 would seem to be fashioned rather after the model of a Bach cantata, and Webern himself more than once referred to the final movement as a chorale (though it is, in fact, canonic). The new suppleness of expressive diction in the solo vocal lines also indicates such a parentage.

Other novel aspects of the Second Cantata include the subtlety with which the orchestra, rich in different types of wind instrument, is employed. In the two choruses (nos.3 and 6) there is a highly variegated doubling of the voices, and in the bass aria the vocal line is wrapped into a canon, as in the soprano aria of the First Cantata. Elsewhere the orchestra provides a fine tracery of brief phrases and chords around the vocal music. The accompaniment is particularly intricate and varied in colour in the fifth movement, where Webern admitted that 'the question of the *economy of instrumentation* – and some compositional things too – held me up'. He also experienced delay in composing the first movement: 'One reason, probably the main one, was the formal thing: a form emerged that must have lain dormant for a long, long time. ... I get to the point of building up 12-note chords'. The accompanying harmonies throughout this number are principally in six parts – again something

116

new for Webern, whose basic chords are usually of four notes or fewer.

Webern's response to the text, in word-painting and symbolism, is particularly rich in these last three choral compositions and, as in Bach, it can be detected at all levels. The vocal writing is supply sensitive to meaning, and Webern did not deny himself the possibilities of orchestral illustration, most notably in the thunder and lightning of the opening chorus of op.29. Less obviously, the sense of the first words of op.31 no.2 – 'Kept deep down, the innermost life sings in the hive' – is consonant with the piece's construction as a bass aria surrounded by short instrumental fragments. Or again, in *Das Augenlicht*, the opening idea of light flowing into the eyes and out again provides the pretext for a serial retrogradation. The susceptibility of Jone's metaphors to such musical interpretations was one of the aspects that attracted Webern to her verse.

After completing the Second Cantata Webern began planning a three-movement concerto, but he altered this project in favour of another choral work to words by Jone. Work proceeded on this during the summer of 1944 and perhaps thereafter, yet the sketchbook contains only fragmentary ideas.

CHAPTER FIVE

Style and ideas

I Style

It is not easy to define the general characteristics of Webern's style, partly because his output is divided by two or three abrupt changes of emphasis, and partly because the only works to have received much serious analytical attention are the serial instrumental compositions from op.21 to op.28. The major landmarks have already been indicated: the moves from tonality to atonality (1908–9), from instrumental miniatures to songs (1913–14) and from extremes of tension to symmetry and serenity (1927–8). Those broad features which remain constant – brevity, the importance of silence, the usually restrained dynamic range, clarity of texture and simplicity of harmony – do not go far in explaining what makes up Webern's style, yet even they are open to exception: the *pianissimo espressivo* label, for instance, cannot be applied to most of the middle-period songs. Webern's brevity, however, is without exception, exceptional. After the Passacaglia (at about ten minutes his longest movement) he wrote no continuous music playing for more than about five minutes, and most movements are considerably briefer than that; his whole output can be heard in about four hours. (It is puzzling that, despite his unfailing use of metronome markings, Webern habitually overestimated the durations of his pieces, often by a factor of two or more.)

While allowing the variety of Webern's music in most aspects but length, it can be argued that the tendency to think in terms of themes built from short motifs, already noted in the case of the serial instrumental works, can be traced in earlier compositions. The published lectures, as well as the Bach arrangement, indicate that Webern was keenly aware of motivic organization, and the use of recurrent patterns, frequently of three or four notes, is observable quite clearly in the earlier atonal compositions (ex.1 shows some instances). Structuring with

Ex.2
(a) String Quartet (1905)

(b) Six Bagatelles, op.9 (1913); no.5

(c) Five Canons, op.16 (1923–4); no.1

(d) Concerto, op.24 (1933–4); first movt.

pitch-class sets is discussed by Forte (1973), who includes an analysis of op.7 no.3 in these terms. That small sets are important in Webern's music of all periods is shown by ex.2, which displays extracts from the openings of four very diverse pieces, each demonstrating the melodic or harmonic use of the same three-note set, that which in its closest configuration, as in (a), includes intervals of a semitone and a minor 3rd (e.g. C,

C♯, E). This set plays a significant part in the construction of those compositions quoted, and others; and different 'semitone-containing' sets can be found throughout Webern's music (the predominance of intervals of a 7th and a minor 9th is, of course, not unrelated).

Though Webern left few statements about his methods and aims until the serial period, the employment of small sets in earlier works can be seen as indicating a desire to elaborate a whole composition from the smallest unit. The facility with which serialism could be used to bring about such motivic coherence – that coherence (*Zusammenhang*) which runs like an *idée fixe* through his published lectures – was probably the main reason for Webern's enthusiasm for the method. Instead of having the series build up an extended melodic theme (as, for example, in Schoenberg's Orchestral Variations), he divided it into smaller groups, again often of three or four notes; and these motifs are the fundamental structural elements, used to form themes and yet still differentiated (compare Webern's own Orchestral Variations). The resulting benign antagonism between thematic organization and motivic organization has already been alluded to.

Frequently the supply of motifs is limited by symmetries within Webern's series. Thus that of the String Quartet consists of three statements of the same set, which in one form spells B–A–C–H. The monomotivicism of the Concerto is achieved by the employment of a series built from one three-note set in the forms prime, transposed retrograde, transposed retrograde inversion, transposed inversion (ex.2*d* shows the first occurrence of the series, where it is disposed instrumentally into the constituent sets). Webern's last project, the unfinished 'op.32', is based on a similar but even more restrictive series, in which each of the identical quarters is simply a fragment of the chromatic scale. Series which contain repeating units, such as those of opp.24 and 28, obviously limit the types of interval and motif available, but in doing so they make possible the highly integrated quality Webern strove for. In discussing the Symphony he suggested that he had gone further than the Netherlands polyphonists in this direction; indeed, the pervasive motivicism of such a composition as Isaac's *Missa super 'O praeclara'* is not so far from Webern's thought. His interpretation of serialism made possible the creation of works where the smallest items of pitch structure could be related to one or two basic ideas: 'always different and yet always the same'.

In order that the special properties of Webern's series should make themselves felt at the surface, it was clearly necessary for each heard element, melodic or harmonic, to be part or whole of a serial motif, or group of serial motifs. It may be for this reason that Webern returned, after the String Trio, to the type of serial composition he had first used in op.18 no.3: each polyphonic line is given its own sequence of serial forms. A serial statement is thus not, as often in Schoenberg, the binding element between melody and accompaniment; in Webern concurrent voices are frequently linked by a delight in emphasizing the motivic coincidences that are bound to result from the use of his highly redundant series (an example from the Symphony has been referred to). Even in those rare cases where fragments of the same serial form are allowed to overlap, as in ex.2*d*, rhythm, timbre and other aspects draw attention to those motifs proper to the series. After op.20 Webern did not extract non-consecutive melodic ideas from the series in the way that Schoenberg and Berg commonly did. And, also unlike them, he made no returns to non-serial practice. Everything suggests that in serialism he had achieved his ideal, the means to achieve a thorough coherence, the 'law' he so much respected.

If Webern found in serialism the restraint he required on the small scale (and his remarks on the Bagatelles indicate how ill at ease he was in the seemingly unbounded universe of atonality), he provided himself with longer-term limitations by the use of established formal methods and models, to which he returned in the first serial compositions: the Piano Piece of 1925 is cast as a minuet. The sorts of series he employed can be said to relate to the sorts of form he used – a series containing repeating sets, for example, is obviously not inimical to canonic presentation, and palindromic series, such as that of the Symphony, lend themselves to palindromic forms – but it would be an exaggeration to describe Webern's forms as 'self-generating' from their series, or to suggest that he initiated 'serial form'. It was part of his achievement to demonstrate how conventional patterns (fugue, sonata, rondo, variations and so on) could be used in a quite new way through a different understanding of what constitutes a theme. In the works of 1909–14 he had departed far from thematicism and traditional form, but his reinterpretations of these in the works of 1927–43 are no less radical.

Generally, however, it is Webern's form, as opposed (if it can be) to his serial structuring, that has been least influential; and influential his music certainly was in the two decades after his death. During his lifetime his work made little impression on composers outside the circle of Schoenberg's pupils (Dallapiccola was one of the few important exceptions), but after the war there was an extraordinary upsurge of interest in his music. Young composers found Webern 'THE threshold' (Boulez), giving a much inferior place to Schoenberg; a piece such as Stockhausen's *Kontra-Punkte* serves to show how much – and how little – they took from Webern. What they discovered in him was an uncompromising rigour in serial usage, a complete break with forms from the past, serial organizations of rhythm, timbre and dynamics, and a union of horizontal and vertical writing. It was not the first instance of creative misunderstanding. In the USA Webern was valued by theorists and composers, Babbitt and Perle among them, for his contributions to serial method; while others, such as Cage and Feldman, were attracted by the quiet and tranquillity of the early atonal music. Yet perhaps the most faithful 'post-Webernian' turned out to be Stravinsky.

II **Ideas**

During the period of Webern's study with Schoenberg, two of the thinkers who exerted the greatest influence on Viennese culture were Maeterlinck and Strindberg, both of whom were seen as attempting, through symbolism and occultism, to penetrate to unspoken levels of thought. Their influence on Webern is attested by his projected and accomplished settings of writings by them and others (Trakl was as indebted to Strindberg as George was to the French symbolists), by the prominence of their works in his library and by such

*10. Anton
Webern*

hints as are provided by the correspondence. He was perhaps most explicit in a letter of 1911 to Berg, drawing certain comparisons: 'And Strindberg and Mahler? Maeterlinck and Schoenberg? Also Strindberg and Schoenberg! Rays of God'.

For Webern, however, the most important teacher was the world of nature, which he had learnt to appreciate as a boy in the Carinthian Alps; knowledge of God was to be had through the study of creation. As he wrote in 1919, again to Berg: 'I love all nature, but, most of all, that which is found in the mountains. For a start I want to progress in the purely physical knowledge of all these phenomena. . . . Experimenting, observing in physical nature is the highest metaphysical theosophy to me'. His most prized books were the repositories for pressed flowers, and it would be hard to judge whether the texts or the specimens meant more to him. The writer who was probably the most significant influence on his ideas in later years was Goethe, the Goethe of the late nature poetry (set in opp.12 and 19), the *Farbenlehre* and *Die Metamorphose der Pflanzen*. Here he found an approach to nature akin to his own: at once analytical and contemplative, attentive to the tiniest detail and yet searching for the coincidence, the general principle.

Moreover, in his serial composition, Webern applied this method in reverse. He regarded the series, or even the sub-serial motif, as a natural law, its operation giving a fundamental coherence to the most varied phenomena, just as Goethe had recognized that all flowering plants, however different their appearances, were unified by the nature of their reproductive cycle. Webern came particularly close to Goethean terminology in writing of the Orchestral Variations to Jone: 'six notes are given, in a *shape* determined by the sequence and the rhythm, and what follows . . . is nothing other than this shape over and over again!!!'. He went on to quote Goethe on the 'prime phenomenon': 'ideal as the ultimate recognizable thing, real when recognized, symbolic, since it embraces every case, identical with every case'.

This suggests a view that Webern touched on in *Der Weg zur neuen Musik*, namely that 'between the products of nature and those of art no essential difference prevails'. Seeing the compositional process as analogous to nature's development of diversity from general principle, he intended the finished work to match those natural phenomena he most admired in clarity, colour and formal perfection. There are hints of this in a programme he wrote for the projected concerto of 1928, where against 'Rondo' he marked: '(Dachstein, snow and ice, crystal-clear air, cosy, warm, sphere of the high pastures) – coolness of the first spring (Anninger, first flora, primroses, anemones (hepatica, pulsatilla))'. And the words of Goethe which Webern quoted in connection with the Second Cantata are indeed as appropriate to his music as to the unity of multiplicity in plant life: 'All shapes are similar and none are the same; thus the chorus points to a secret law, to a holy riddle'.

WORKS

This list includes all published works, together with those unpublished works which had been performed by January 1979. Moldenhauer (1978) lists many other unpublished works, all predating op.1, as well as numerous sketches. The only sketches listed here are those published in *A. von Webern: Sketches (1926–1945)* (New York, 1968) [S].

Numbers in the right-hand column denote references in the text.

ORCHESTRAL

op.		
—	Im Sommerwind, idyll, after B. Wille, 1904	
—	Three Studies on a Ground, 1907, unpubd; related to op.11	
1	Passacaglia, 1908; arr. 2 pf 6 hands, 1918, lost	90, 93
5	Five Movements, str, arr. from str qt 1928, rev. 1929	97
6	Six Pieces, 1909; arr. fl, ob, cl, harmonium, pf, perc, str qt, 1920, unpubd; arr. reduced orch, 1928	96, 97
10	Five Pieces, small orch, 1911–13; arr. harmonium, pf qt, 1919, unpubd	91, 98, 99
—	Five Pieces, 1913; related to opp.6 and 10	
—	Eight Fragments, 1911–13, unpubd; related to op.10	
21	Symphony, 1928; projected 3rd movt, S	107, 109, 110, 111, 122
—	Concerto, vn, cl, hn, pf, str, 1928, S; reconceived as Quartet, op.22	92, 109, 111, 120, 125
—	Overture, 1931, S	117
30	Variations, 1940	

CHORAL

—	Concerto, 1944, S; reconceived as Cantata [no.3]	
2	Entflieht auf leichten Kähnen (George), SATB, 1908; acc. for harmonium, pf qt, 1914, unpubd	94
19	Two Songs (Goethe), SATB, cl, b cl, cel, gui, vn, 1926, vocal score, 1928: Weiss wie Lilien, Zieh'n die Schafe von der Wiese; projected 3rd no. Auf Bergen in der reinsten Höhe (Goethe), SATB, 1927, S	107
—	Der Spiegel sagt mir (Goethe), SSAA, 1930, S	
—	Wie kann der Tod so nah der Liebe wohnen? (Jone), SATB, acc., 1934, S; related to Concerto, op.24	
26	Das Augenlicht (Jone), SATB, orch, 1935	115, 117
29	Cantata no.1 (Jone), S, SATB, orch, 1938–9; vocal score, 1944	115
31	Cantata no.2 (Jone), S, B, SATB, orch, 1941–3; vocal score, 1944; projected movt Kleiner sind Götter geworden (Jone), 1943–4, S	115, 116, 126
—	Cantata [no.3] (Jone), section Das Sonnenlicht spricht, 1944, S	

SOLO VOCAL

—	Wolkennacht (Avenarius), 1v, pf, 1900, unpubd	
—	Vorfrühling II (Avenarius), 1v, pf, 1900, unpubd	
—	Wehmut (Avenarius), 1v, pf, 1901, unpubd	
—	Three Poems, 1v, pf, 1899–1903: Vorfrühling (Avenarius), Nachtgebet der Braut (Dehmel), Fromm (Falke)	
—	Two Songs (Avenarius), 1v, pf, 1900–01, unpubd: Wolkennacht, Wehmut	
—	Hochsommernacht (Greif), S, T, pf, 1904, unpubd	
—	Eight Early Songs, 1v, pf, 1901–4: Tief von fern (Dehmel), Aufblick(Dehmel), Blumengruss(Goethe), Bild der Liebe(Greif), Sommerabend (Weigand), Heiter (Nietzsche), Der Tod (Claudius), Heimgang in der Frühe (von Liliencron)	
—	Siegfrieds Schwert (Uhland), 1v, orch, 1903, unpubd	
—	Three Songs (Avenarius), 1v, pf, 1903–4; Gefunden, Gebet, Freunde	
—	Five Songs (Dehmel), 1v, pf, 1906–8: Ideale Landschaft, Am Ufer, Himmelfahrt, Nächtliche Scheu, Helle Nacht	94
3	Five Songs from 'Der siebente Ring' (George), 1v, pf, 1908–9: Dies ist ein Lied, Im Windesweben, An Bachesranft, Im Morgentaun, Karl reckt der Baum	94, 95
4	Five Songs (George), 1v, pf, 1908–9: Eingang, Noch zwingt mich Treue, Heil und Dank dir, So ich traurig bin, Ihr tratet zu dem Herde	94, 95
—	Four Songs (George), 1v, pf, 1908–9: Erwachen aus dem tiefsten Traumesschosse, Kunfttag I, Trauer I, Das lockere Saatgefilde lechzet krank	
8	Two Songs (Rilke), Mez, cl + bcl, hn, tpt, cel, harp, vn, va, vc, 1910, vocal score, 1925, unpubd: Du, der ichs nicht sage, Du machst mich allein	97
—	Schmerz, immer blick' nach oben (Webern), 1v, str qt, 1913, unpubd; associated with Six Bagatelles, op.9	100

— Three Songs, S, small orch: Leise Düfte (Weber), 1914; Kunfttag III (George), 1914; O sanftes Glühn der Berge (Weber), 1913; no.2 reconstructed by P. Westergaard from sketch

12 Four Songs, 1v, pf: Der Tag ist vergangen (trad.), 1915; Die geheimnisvolle Flöte (Li Tai Po, trans. Bethge), 1917; Schien mir's, als ich sah die Sonne (Strindberg, Ger. trans.), 1915; Gleich und gleich (Goethe), 1917 — 102

13 Four Songs, S, small orch: Wiese im Park (Kraus), 1917; Die Einsame (Wang Seng Yu, trans. Bethge), 1914; In der Fremde (Li Tai Po, trans. Bethge), 1917; Ein Winterabend (Trakl), 1918; vocal score, 1924 — 102, 103

14 Six Songs (Trakl), S, cl + Eb-cl, b cl, vn, vc: Die Sonne, 1921; Abendland I, 1919; Abendland II, 1919; Abendland III, 1917; Nachts, 1919; Gesang einer gefangenen Amsel, 1919; vocal score, 1923, unpubd — 103

15 Five Sacred Songs, S, fl, cl + b cl, tpt, harp, vn + va: Das Kreuz, das musst' er tragen (trad.), 1921; Morgenlied (Des Knaben Wunderhorn), 1922; In Gottes Namen aufstehen (trad.), 1921; Mein Weg geht jetzt vorüber (trad.), 1922; Fahr hin, o Seel', zu deinem Gott (trad.), 1917; vocal score, 1923, unpubd — 104

16 Five Canons on Latin Texts, S, cl, b cl, 1923-4: Christus factus est (Maundy Thursday liturgy); Dormi Jesu (Des Knaben Wunderhorn), Crux fidelis (Good Friday liturgy), Asperges me (Maundy Thursday liturgy), Crucem tuam adoramus (Good Friday liturgy) — 105

17 Three Traditional Rhymes, S, cl, b cl, vn + va, 1924-5: Armer Sünder, du; Liebste Jungfrau, wir sind dein; Heiland, unsre Missetaten — 106

18 Three Songs, S, Eb-cl, gui, 1925: Schatzerl klein, musst nit traurig sein (trad.); Erlösung (Des Knaben Wunderhorn); Ave regina coelorum — 106

— Nun weiss man (Goethe), 1v, pf, 1929, S

— Doch immer höher steigt der edle Drang', 1v, pf, 1930, S

— Der Spiegel sagt mir (Goethe), 1v, insts, 1930, S

23 Three Songs from 'Viae inviae' (Jone), 1v, pf, 1933-4: Das dunkle Herz, Es stürzt aus Höhen Frische, Herr Jesus mein — 114

25 Three Songs (Jone), 1v, pf, 1934: Wie bin ich froh, Des Herzens Purpurvogel fliegt durch Nacht, Sterne, ihr silbernen Bienen — 114

CHAMBER AND INSTRUMENTAL

— Two Pieces, vc, pf, 1899

— Scherzo and Trio, a, str qt, 1904, unpubd

— Slow Movement, str qt, 1905 — 93, 94

— String Quartet, 1905

— Movement, pf, 1906

— Rondo, str qt, 1906

— Sonata Movement, pf, 1906 — 93

— Piano Quintet, 1907

— String Quartet, a, 1907, unpubd

5 Five Movements, str qt, 1909 — 96

7 Four Pieces, vn, pf, 1910 — 96, 119

9 Six Bagatelles, str qt, 1911-13 — 98, 122

11 Three Little Pieces, vc, pf, 1914 — 98, 101

— Sonata, vc, pf, 1914

— Trio Movement, cl, tpt, vn, 1920, unpubd — 101

— Kinderstück, pf, 1924 — 105

— Piano Piece, '1925 — 105, 122

— Movement ('Ruhig fliessend'), str trio, 1925 — 105

— Movement ('Ruhig'), str trio, 1925, unpubd

20 String Trio, 1926-7 — 106, 108

— Movement ('Sehr lebhaft'), str trio, 1927, unpubd; orig. intended for op.20

22 String Quartet, 1929, S — 109, 112, 113

24 Concerto, cl, t sax, pf, vn, 1930; projected 3rd movt, 1930, S — 109, 110

— Quartet, cl, ob, cl, hn, tpt, trbn, pf, vn, va, 1931-4; abandoned movt, 1934, S; sketches with different instrumentations, 1934, S

27 Piano Variations, 1935-6 — 109, 111

28 String Quartet, 1936-8 — 109, 112, 121

ARRANGEMENTS AND EDITIONS

F. Schubert: Thränenregen, Ihr Bild, Romance from 'Rosamunde', Der Wegweiser, Du bist die Ruh', orchd 1903, unpubd; Deutsche Tänze D.820, orchd 1931

H. Isaac: Choralis constantinus II, DTÖ, xxxii, Jg.xvi/1 (1909)

A. Schoenberg: Prelude and Interludes from 'Gurrelieder', 2 pf 8 hands, 1909-10, unpubd; Six Songs op.8, vocal score, 1910; Five Pieces op.16, 2 pf 4 hands, 1912; Chamber Symphony op.9, fl/vn, cl/va, pf, vn, vc, 1922-3

J. Strauss II: *Schatzwalzer*, pf qnt, harmonium, 1921, unpubd
F. Liszt: *Arbeiterchor*, B, chorus, orch, 1924, unpubd; vocal score, 1924

J. S. Bach: *Fuga (Ricercata) a 6 voci*, orchd 1934–5   111
R. Wagner-Régeny: *Johanna Balk*, vocal score, 1939
O. Schoeck: *Das Schloss Dürande*, vocal score, 1941–2
Principal publishers: Fischer, Universal
MSS in the Music Library, Northwestern University, Evanston, Illinois

WRITINGS

'Der Lehrer', *Arnold Schoenberg* (Munich, 1912), 85; Eng. trans. in Wildgans (1966), 160

'Schoenbergs Musik', *Arnold Schoenberg* (Munich, 1912), 22
Tot: sechs Bilder für die Bühne (in memoriam ... Oktober 1913), 100 unpubd
'Aus Schoenbergs Schriften', *Arnold Schoenberg zum 60. Geburtstag* (Vienna, 1934)
ed. W. Reich: *Der Weg zur neuen Musik* (Vienna, 1960; Eng. trans., 92, 125 1963)
——: *Anton Webern: Weg und Gestalt: Selbstzeugnisse und Worte der Freunde* (Zurich, 1961)
'Youthful Poems' in Wildgans (1966), 163

MSS in University of Washington Music Library, Seattle

BIBLIOGRAPHY

CORRESPONDENCE
'Letters from Webern and Schoenberg to Roberto Gerhard', *Score* (1958), no.24, p.36

J. Polnauer, ed.: *Briefe an Hildegard Jone und Josef Humplik* (Vienna, 1959; Eng. trans., 1967)

W. Reich, ed.: 'Briefe aus Weberns letzten Jahren', *ÖMz*, xx (1965), 407

I. Vojtéch, ed.: 'Arnold Schönberg, Anton Webern, Alban Berg – unbekannte Briefe an Ervin Schulhoff aus den Jahren 1919–26', *MMC*, xviii (1965), 31–83

W. Reich, ed.: 'Berg und Webern schreiben an Hermann Scherchen', *Melos*, xxxiii (1966), 225

H. Lindlar, ed.: 'Briefe der Freundschaft (1911–1945)', *Kontrapunkte*, ii (1968), 126

BIBLIOGRAPHIES
A. P. Basart: *Serial Music: a Classified Bibliography of Writings on Twelve-tone and Electronic Music* (Berkeley and Los Angeles, 1961)

M. Fink: 'Anton Webern: Supplement to a Basic Bibliography', *CMc* (1973), no.16, p.103

K. Thompson: *A Dictionary of Twentieth-century Composers 1911–1971* (London, 1973)

See also Moldenhauer and Irvine (1966) and Kolneder (Eng. trans., 1968)

MONOGRAPHS AND COLLECTIONS OF ESSAYS
'Webern zum 50. Geburtstag', *23* (Vienna, 1934), no.14

'Anton Webern: Dokumente, Bekenntnis, Erkenntnisse, Analysen', *Die Reihe* (1955), no.2; Eng. trans., *Die Reihe* (1958), no.2

W. Kolneder: *Anton Webern: Einführung in Werk und Stil* (Rodenkirchen, 1961; Eng. trans., 1968)

H. Moldenhauer: *The Death of Anton Webern: a Drama in Documents* (New York, 1961)

H. Moldenhauer and D. Irvine, eds.: *Anton von Webern: Perspectives* (Seattle, 1966)

F. Wildgans: *Anton Webern* (London, 1966)

R. V. Ringger: *Anton Weberns Klavierlieder* (Zurich, 1968)

L. Somfai: *Anton Webern* (Budapest, 1968)

C. Rostand: *Anton Webern: l'homme et son oeuvre* (Paris, 1969)

D. Beckmann: *Sprache und Musik im Vokalwerk Anton Weberns*, Kölner Beiträge zur Musikforschung, lvii (Regensburg, 1970)

'Anton von Webern', *ÖMz*, xxvii/3 (1972)

5. Internationaler Webern-Kongress: Wien 1972

Bibliography

W. Kolneder: *Anton Webern: Genesis und Metamorphose eines Stils* (Vienna, 1974)

F. Döhl: *Weberns Beitrag zur Stilwende der neuen Musik* (Munich, 1976)

H. Moldenhauer: *Anton von Webern: Chronicle of his Life and Works* (New York and London, 1978; Ger. trans., 1979)

OTHER GENERAL LITERATURE

H. Searle: 'Conversations with Webern', *MT*, lxxxi (1940), 405

R. Leibowitz: *Schönberg et son école* (Paris, 1947; Eng. trans., 1949)

L. Rognoni: *Espressionismo e dodecafonia* (Turin, 1954, rev. 2/1966 as *La scuola musicale di Vienna*)

H. Pousseur: 'Webern und die Theorie', *Darmstädter Beiträge zur neuen Musik*, i (1958), 38

H. Searle: 'Studying with Webern', *RCM Magazine* (1958), no.2, p.39; Swed. trans. in *Nutida musik*, ii/4 (1958–9), 11

F. D. Dorian: 'Webern als Lehrer', *Melos*, xxvii (1960), 101

P. Stadlen: 'The Webern Legend', *MT*, ci (1960), 695

P. Boulez: 'Webern, Anton von', *Encyclopédie de la musique*, iii (Paris, 1961); repr. in *Relevés d'apprenti* (Paris, 1966; Eng. trans., 1968; It. trans., 1968), 367

G. Perle: *Serial Composition and Atonality: an Introduction to the Music of Schoenberg, Berg and Webern* (Berkeley and Los Angeles, 1962, 4/1975)

G. Rochberg: 'Webern's Search for Harmonic Identity', *JMT*, vi (1962), 109

F. Döhl: 'Die Welt der Dichtung in Weberns Musik', *Melos*, xxxi (1964), 88

G. Ligeti: 'Weberns Melodik', *Melos*, xxxiii (1966), 116

R. U. Ringger: 'Sprach-musikalische Chiffern in Anton Webern's Klavierliedern', *SMz*, cvi (1966), 14

L. Stein: 'The Privataufführungen Revisited', *Paul A. Pisk: Essays in his Honor* (Austin, 1966), 203

R. U. Nelson: 'Webern's Path to the Serial Variation', *PNM*, vii/2 (1969), 73

G. Perle: 'Webern's Twelve-tone Sketches', *MQ*, lvii (1971), 1

U. von Rauchhaupt: *Die Streichquartette der Wiener Schule: Schoenberg Berg Webern* (Munich, 1971; Eng. trans., 1971)

P. K. Bracanin: 'The Palindrome: its Application in the Music of Anton Webern', *MMA*, vi (1972), 38

K. Bailey: 'The Evolution of Variation Form in the Music of Webern', *CMc* (1973), no.16, p.55

A. Forte: *The Structure of Atonal Music* (New Haven and London, 1973)

Webern

LITERATURE ON SPECIFIC WORKS
(*tonal works*)

E. T. Cone: 'Webern's Apprenticeship', *MQ*, liii (1967), 39

R. Gerlach: 'Mystik und Klangmagie in Anton von Weberns hybrider Tonalität', *AMw*, xxxiii (1976), 1

See also Moldenhauer and Irvine (1966)

(*atonal works*)

E. Stein: 'Anton Webern, Fünf Stücke für Orchester', *Pult und Taktstock*, iii (1926), 109

R. U. Ringger: 'Zur Wort-Ton-Beziehung beim frühen Anton Webern: Analyse von Op.3, Nr.1', *SMz*, ciii (1963), 330

J. Baur: 'Über Anton Weberns Bagatellen für Streichquartett', *Veröffentlichungen des Instituts für neue Musik und Musikerziehung Darmstadt*, vi (1967), 62

A. Elston and others: 'Some Views of Webern's Op.6, No.1', *PNM*, vi/1 (1967), 63

H. Kaufmann: 'Figur in Weberns erster Bagatelle', *Veröffentlichungen des Instituts für neue Musik und Musikerziehung Darmstadt*, vi (1967), 69

J. Hansberger: 'Anton Webern: die vierte Bagatelle für Streichquartett als Gegenstand einer Übung im Musikhören', *Musica*, xxiii (1969), 236

H.-P. Raiss: 'Analyse der Bagatelle op.9, no.5 von Anton Webern', *Veröffentlichungen des Instituts für neue Musik und Musikerziehung Darmstadt*, viii (1969), 50

E. Budde: *Anton Weberns Lieder, op.3: Untersuchungen zur frühen Atonalität bei Webern* (Wiesbaden, 1971)

B. Archibald: 'Some Thoughts on Symmetry in Early Webern: Op.5, No.2', *PNM*, x/2 (1972), 159

S. Persky: 'A Discussion of Compositional Choices in Webern's *Fünf Sätze für Streichquartett*, Op.5, First Movement', *CMc* (1972), no.13, p.68

P. Westergaard: 'On the Problems of "Reconstruction from a Sketch": Webern's *Kunfttag III* and *Leise Düfte*', *PNM*, xi/2 (1973), 104

R. Travis and A. Forte: 'Analysis Symposium: Webern, Orchestral Pieces (1913), Movement 1 ("Bewegt")', *JMT*, xviii (1974), 2 [see also xix (1975), 47]

T. Olah: 'Weberns vorserielle Tonsystem', *Melos/NZM*, i (1975), 10

C. Wintle: 'An Early Version of Derivation: Webern's op.11/3', *PNM*, xiii/2 (1975), 166

R. Chrisman: 'Anton Webern's "Six Bagatelles for String Quartet", op.9: the Unfolding of Intervallic Successions', *JMT*, xxiii (1979), 81–122

132

Bibliography

(serial instrumental works)

E. Stein: 'Weberns Trio op.20', *Neue Musikzeitung* (1928), 517

——: 'Webern's New Quartet', *Tempo* (1939), no.4, p.6

R. Leibowitz: *Qu'est ce que la musique de douze sons?: le Concerto pour neuf instruments, op.24, d'Anton Webern* (Liège, 1948)

K. Stockhausen: 'Weberns Konzert für 9 Instrumente, op.24: Analyse des ersten Satzes', *Melos*, xx (1953), 343

A. Elston: 'Some Rhythmic Practices in Contemporary Music', *MQ*, xlii (1956), 318 [incl. discussion of op.22]

C. Mason: 'Webern's Later Chamber Music', *ML*, xxxviii (1957), 232

W. F. Goebel: 'Weberns Sinfonie', *Melos*, xxviii (1961), 359

D. Lewin: 'A Metrical Problem in Webern's Op.27', *JMT*, vi (1962), 124

W. L. Ogdon: 'A Webern Analysis', *JMT*, vi (1962), 133 [on op.27]

P. Westergaard: 'Some Problems in Rhythmic Theory and Analysis', *PNM*, i/1 (1962), 180 [on op.27]

F. Döhl: 'Weberns Opus 27', *Melos*, xxx (1963), 400

P. Westergaard: 'Webern and "Total Organization": an Analysis of the Second Movement of the Piano Variations, Op.27', *PNM*, i/2 (1963), 107

S. Borris: 'Structural Analysis of Webern's Symphony, Op.21', *Paul A. Pisk: Essays in his Honor* (Austin, 1966), 231

B. Fennelly: 'Structure and Process in Webern's Opus 22', *JMT*, x (1966), 300

H. Deppert: 'Rhythmische Reihentechnik in Weberns Orchestervariationen Op.30', *Festschrift Karl Marx zum 70. Geburtstag* (Stuttgart, 1967), 84

L. Hiller and R. Fuller: 'Structure and Information in Webern's Symphonie, Op.21', *JMT*, xi (1967), 60–115

S. Goldthwaite: 'Historical Awareness in Anton Webern's *Symphony, Op.21*', *Essays in Musicology: in Honor of Dragan Plamenac* (Pittsburgh, 1969), 65

M. Boykan: 'The Webern Concerto Revisited', *Proceedings of the American Society of University Composers*, iii (1970), 74

M. Starr: 'Webern's Palindrome', *PNM*, viii/2 (1970), 127 [on op.21]

H. Deppert: *Studien zur Kompositionstechnik im instrumentalen Spätwerk Anton Weberns* (Darmstadt, 1972)

J. W. Reid: 'Properties of the Set Explored in Webern's Variations, op.30', *PNM*, xii/1–2 (1973–4), 344

D. Cohen: 'Webern and the Magic Square', *PNM*, xiii/1 (1974), 213

R. Smalley: 'Webern's Sketches', *Tempo* (1975), no.112, p.2

C. Wintle: 'Analysis and Performance: Webern's Concerto op.24/II', *Music Analysis*, i/1 (1982), 73

(serial vocal works)

N. Castiglioni: 'Sul rapporto tra parola e musica nella Seconda Cantata di Webern', *Incontri musicali* (1959), no.3, p.112

G. Ligeti: 'Über die Harmonik in Weberns erste Kantate', *Darmstädter Beiträge zur neuen Musik*, iii (1960), 49

L. Spinner: 'Anton Weberns Kantate Nr.2, Opus 31: die Formprinzipien der kanonischen Darstellung', *SMz*, ci (1961), 303

E. Klemm: 'Symmetrien im Chorsatz von Anton Webern', *DJbM*, ix (1966), 107

R. U. Ringger: 'Reihenelemente in Anton Weberns Klavierlieder', *SMz*, cvii (1967), 144

D. Saturen: 'Symmetrical Relationships in Webern's First Cantata', *PNM*, vi/1 (1967), 142

H. Moldenhauer: 'Webern's Projected Op.32', *MT*, cxi (1970), 789

D. Chittum: 'Some Observations on the Row Techniques in Webern's *Opus 25*', *CMc* (1971), no.12, p.96

P. Luckman: 'The Sound of Symmetry: a Study of the Sixth Movement of Webern's *Second Cantata*', *MR*, xxxvi (1975), 187

R. L. Todd: 'The Genesis of Webern's opus 32', *MQ*, lxvi (1980), 581

(arrangements)

C. Dahlhaus: 'Analytische Instrumentation', *Bach-Interpretationen: Walter Blankenburg zum 65. Geburtstag* (Göttingen, 1969), 197

BERG

George Perle

CHAPTER ONE

Early years and first works

Alban Maria Johannes Berg was born in Vienna on 9 February 1885. He lived there all his life, but spent some months of every year in the Carinthian Alps, first at the family estate, Berghof, near Villach, later also at the home of his wife's family in Trahütten, and finally in the Waldhaus on Lake Wörther, which he purchased, in spite of his straitened financial circumstances at the time, just three years before his death.

His musical education before his meeting with Schoenberg in October 1904 was negligible, to judge from the numerous songs that he had composed by that time. Berg looked upon Schoenberg not only as his composition teacher, but as his model and mentor in all things and as a surrogate for his father, who had died when Berg was 15. His formal education had come to an end a few months before he met Schoenberg, when he finally passed his school examinations after having failed them in the previous year. But the Gymnasium in Vienna, as Berg's friend Stefan Zweig described it in *The World of Yesterday* (New York, 1943), was 'monotonous, heartless and lifeless', where 'we heard nothing new or nothing that to us seemed worth knowing', while 'outside there was a city of a thousand attractions, a city with theatres, museums, bookstores, universities, music'. The complacent conservatism of Viennese culture was being challenged in every field by men who were not yet illustrious but who were already familiar to young men like Zweig and Berg. Kraus, Klimt, Loos, Altenberg, Kokoschka – such men belonged to the young Berg's intimate circle, as is shown by letters to his future wife in which he described their coffee-house encounters and midnight cabaret gatherings.

Berg's association with Schoenberg, momentous as it was, resulted from a fortuitous circumstance, the name having come to his attention only through an advertisement for students that Schoenberg had placed in a newspaper. Schoenberg, 30 years of age when Berg became his pupil, had only recently completed *Pelleas und Melisande*; and his first large work, *Verklärte Nacht*, though composed in 1899, had only just had its first performance. Berg and Webern, whose lessons with Schoenberg started at about the same time, were thus to share with their master the creative experiences that led him through a series of radical stylistic changes, from the First String Quartet of 1904–5 to the first expressionist works, above all the Five Orchestral Pieces and *Erwartung*, both composed in 1909.

By this time Berg was nearing the end of his apprenticeship. The Seven Early Songs (1905–8), which he revised and published in 1928, are still in the tradition of the Romantic German lied and show influences that range from Schumann to early Schoenberg. The one-movement Piano Sonata op.1 (1907–8) recapitulates on a miniature scale the principal features of the master's more advanced tonal works of 1904–8. The complex interplay and transformation of a few motivic units generate both melodic and accompanying material of the larger thematic components. Clear tonal resolutions

*11. Portrait
(c1910) of
Berg by
Arnold
Schoenberg*

139

at salient structural points give direction to the work as
a whole in spite of the high degree of chromaticism and
tonal ambiguity. In Berg's next work, the Four Songs
op.2, the traditional concept of tonal centricity is
already radically questioned. Presumably the signature
of six flats in the second song is intended to imply the
tonality of E♭ minor; but the piece begins and ends on
chords that are the enharmonic equivalent of the French
6th, and the piece may be analysed almost entirely as a
chain of these chords. The semitonal shift of such a
chord produces the same collection of pitch classes as
its transposition by a perfect 5th (ex.1), and the piece is

Ex.1

more easily explained by the implications of this
property than by the traditional tonal functions. The
final song of op.2 is Berg's first definitively 'atonal'
piece, and in it there are already numerous details that
are hallmarks of his personal language. With the String
Quartet op.3 (1910), the last work he was to write under
Schoenberg's tutelage, Berg took his place as an innovatory figure in his own right, a colleague, not merely
a disciple, of Schoenberg in the evolution of the new
music, as Webern had done in the previous year with his
Five Movements for string quartet.

The quartet is remarkably original in every way: in
its brilliant and imaginative expansion of the idiomatic
sonic resources of the medium; in its motivic and
thematic workmanship, which derives from the

140

procedures on which the Piano Sonata is based but brings these to an unprecedented degree of complexity in the absence of the tonal controls that had ultimately governed their use in the sonata; and most significantly in its early use of referential patterns that can function as normative elements in a non-tonal idiom. The opening bars (ex.2) provide a number of striking examples of

Ex.2

such patterns. The head-motif (*a*), which is a pervasive figure throughout the first movement and returns at the close of the quartet, consists of a five-note segment of the whole-tone scale plus one odd note. Such collections, in which one or two elements are displaced in what would otherwise be a complete representation of the whole-tone cycle, are assigned an important structural role in other works, above all in *Wozzeck*. The semitone produced by this displacement is symmetrically expanded (*b*) along inversionally related segments of the semitonal cycle. In the lower voices (*c*) a segment of the semitonal cycle is aligned with a segment of the perfect

4th cycle. Symmetrical inflections (*d*) prolong this progression. An abstract rhythmic pattern is introduced at (*c*), the first dyad being stated once, the second twice, and the third three times. Such devices remained characteristic of Berg's work to the end of his life.

CHAPTER TWO

Opp. 4–6 and World War I

In May 1911 Berg, having won her father's reluctant agreement, married Helene Nahowski, whom he had met, and immediately fallen in love with, four years earlier. In the months following his marriage he worked for Universal Edition, preparing the piano reduction of Schreker's *Der ferne Klang* and Schoenberg's *Gurrelieder*. The String Quartet was first performed on 24 April 1911. It was unsuccessful, not surprisingly in view of the fact that the performance was assigned to an ad hoc ensemble that could hardly have been prepared to cope with this difficult and innovatory work, and it was not heard again until 12 years later when it was revived to great acclaim at a concert of the ISCM.

Lacking self-confidence because of his late start and his incompetence in the 'applied' aspects, performance and conducting, of his art, and not yet given a fair opportunity to judge the effect of his work in performance, Berg remained utterly dependent upon Schoenberg's estimate of his achievement and progress. The 'String Quartet', wrote Schoenberg, shortly after Berg's death,

surprised me in the most unbelievable way by the fullness and unconstraint of its musical language, the strength and sureness of its presentation, its careful working and significant originality. That was the time when I moved to Berlin (1911) and he was left to his own devices. He has shown that he was equal to the task.

At the time, however, the indications that Berg could now be 'left to his own devices' must have created some psychological and emotional difficulties for the master. The first work of Berg's that was not written directly under Schoenberg's supervision led to the first major crisis in their relationship. This was the *Fünf Orchesterlieder nach Ansichtskartentexten von Peter Altenberg*, which Berg completed a year after Schoenberg's move to Berlin. On 31 March 1913 Schoenberg conducted a concert in Vienna which included two of the Altenberg songs. Within the audience was a hostile faction bent on provoking a disturbance, which succeeded in creating such a tumult during the Berg pieces that the concert could not be continued.

Two months later Berg visited Schoenberg in Berlin. From his subsequent letters to Schoenberg it is evident that Schoenberg not only did nothing to reassure Berg, but even added his own censure to that of the critics and the public. According to Berg's 'authorized' biographer, Willi Reich, 'it must have been the aphoristic form of the latest pieces – the Altenberg songs and the Four Pieces for clarinet and piano op.5, completed in the spring of 1913 – that occasioned Schoenberg's vehement censure; they were so brief as to exclude any possibility of extended thematic development'. It is difficult to understand why Schoenberg should have found cause in this to admonish Berg, rather than Webern, whose compositions were invariably 'so brief as to exclude any possibility of extended thematic development'. Moreover, Schoenberg himself had just turned to 'aphoristic form' in his Six Little Piano Pieces and in *Pierrot lunaire*. It is even more difficult to understand why Schoenberg should have performed Berg's op.4 in a fragmentary version that in itself fatally misrepresented the work, in view of his absolutely uncompromising insistence on authenticity in the performance of his own compositions. The experience for Berg was a traumatic one; he never again attempted to bring this, perhaps his finest composition after the two operas and the Lyric Suite, to performance, and it remained unheard and unpublished until 17 years after his death.

Apart from their brevity the Altenberg songs have little in common with the aphoristic statements of Webern. They are miniatures rather, compressing within their very circumscribed durational limits the timbral resources of a very large orchestra and an extraordinarily complex and extensive system of motivic interrelationships. In the *pianississimo* opening bars six simultaneous ostinato figures, no two of them coinciding in length, create a dense mist of sound. Rigorous transformation patterns, analogous to those which open the String Quartet but on a much larger scale, bring order to this chaos by imposing a progression that rises to a *fortissimo* statement of the basic harmonic figure of the work (ex.3), whereupon all is reduced to a quiet background for the first entrance of the solo voice. The effect

Ex.3

is that of a heavy curtain of sound that opens to reveal and frame what is central to the work, articulate song. The effectiveness of the musical imagery and the technical means employed already point to the composer of *Wozzeck*.

A linear version of the first chord of ex.3 (ex.4) is the principal melodic motif of the work. Its principal theme, as opposed to such cell-like motifs, is the earliest example of an ordered 12-note series. The final movement, which closes with a reversed statement of ex.3, is a passacaglia based on three subjects, one of which is this 12-note series. It thus anticipates the oft-cited passacaglia based on a 12-note theme in Act 1 scene iv of *Wozzeck*. Berg's early interest in systematic statements of all 12 pitch classes is also seen in the 12-note chord which opens and closes the third song, where it serves as a musical representation of the word 'All' in the line that opens and, with its verb-tense changed, closes the text: 'Über die Grenzen des All blicktest [blickst] du [noch] sinnend hinaus!' ('Beyond the bounds of the universe musingly you looked [still look]!').

Ex.4

Innovatory and original as is Berg's op.4, its sources in the music of Schoenberg – above all, the third and fourth movements of the Second Quartet, the vocal score of which Berg had arranged for Universal Edition, and the Five Pieces for Orchestra – are unmistakable. Schoenberg's tonal and early atonal works remained an important and continuing influence on Berg's music to the end of his life. Both *Wozzeck* and *Lulu* find their precedent in *Gurrelieder* for the integration, within a continuous large-scale design, of individual numbers that remain structurally self-contained in spite of thematic and harmonic interrelations and cross-references. A procedure whose only precedents are found in two works of Schoenberg's tonal period, the First Quartet and the First Chamber Symphony, plays a most important role in *Lulu*: the separation of sectional components of a single well-defined form (respectively a sonata allegro in Act 1, a rondo in Act 2, and a set of variations in Act 3) through the interposition of sections that are not components of that form.

Though op.5, the Four Pieces for clarinet and piano, may be paired with the Altenberg songs as Berg's only other set of atonal miniatures, it represents a radical departure from its predecessor. Where the Altenberg songs have motifs and themes, the Four Pieces have only cells. What have been here called 'normative' devices in atonal music – repetition, symmetrical chords and progressions, chromatic inflection, progressive transformation patterns – govern the overall shape of each piece, rather than brief episodes only, as in the preceding works. In spite of its microcosmic dimensions (the movements are respectively 12, 9, 18 and 20 bars in length), the Four Pieces have the effect of a large-scale work based on traditional concepts of balance and contrast, in comparison with Webern's Six Bagatelles for string quartet, which were composed in the same year, 1913. Adorno called attention to the analogy with the four-movement Classical sonata, the more complex and extensive outer movements of Berg's miniature cycle corresponding to the sonata allegro and finale of the Classical model, the second to the adagio, and the third to the scherzo. This imposition of a classically proportioned design upon a 'content' of great dramatic power and urgency produces its own tension and strongly characterizes every one of Berg's mature compositions.

Op.5 requires only two performers and lasts only about five minutes, but Berg had to wait more than six years for its first performance. His continuing self-doubt is still painfully evident in a letter to Schoenberg of 20 July 1914, in which he referred to a large work in progress which he hoped would meet with Schoenberg's approval:

But I must always be asking myself whether that which I'm expressing there . . . is any better than the last things I've done. And how can I judge this? Those I hate, so much that I've been close to destroying them, and of these I have no opinion yet, since I'm right in the middle of them.

On 8 September 1914 he sent Schoenberg the first and third movements of the Three Orchestral Pieces op.6, with the hope that this was 'something I could confidently dedicate to you without incurring your displeasure'. Self-criticism and the outbreak of the war ('the urge "to be in it", . . . to serve my country') had slowed up work on the second movement.

If the Three Pieces were to some extent conceived as a project in self-improvement and as a means of winning Schoenberg's approval, they seem also to have been urged upon Berg by the overwhelming impact of Mahler's music, above all the Ninth Symphony, the first performance of which he had heard in June 1912, a year after Mahler's death. By implication at least, the Ninth Symphony has a 'programme', nothing less than the demise of the Classical–Romantic tradition of which it itself is one of the last significant products. In continuing this tradition beyond Mahler, Berg's op.6 is in a sense a retrogressive work, coming as it does after Schoenberg's

Five Orchestral Pieces, *Erwartung* and *Pierrot lunaire*, and after Berg's own Altenberg songs and the clarinet and piano pieces. Like the String Quartet, and even the Piano Sonata, it is concerned with the problem of creating a large-scale musical structure in an ambiguous harmonic language which can no longer give coherence to such a structure, and it attempts to solve this problem through extremely complex thematic operations. The three movements, 'Präludium', 'Reigen' ('Round-dances'), and 'Marsch', are integrated into a single large-scale design through their thematic connections. In this

respect the work bears no resemblance to the three sets of orchestral pieces in which Berg was preceded by Schoenberg (op.16) and Webern (opp.6 and 10). Its complex transformation processes, use of germ motifs and assignment of different functions to variants of the same thematic and motivic elements show Berg still indebted to the pre-atonal works of Schoenberg – above all, perhaps, to the First Quartet, the First Chamber

149

Symphony and even *Pelleas und Melisande*. The two principal themes of the second movement (exx.5 and 6) are first announced in the 'Präludium'. Another important theme of the second movement is derived from the sextuplet figure of ex.5 (ex.7), and one of the principal subjects of the third movement is derived, through inversion and rhythmic variation, from the topmost line of ex.6 (ex.8). Primary thematic components of the whole work thus have their genesis in the first movement.

Ex.7

Ex.8

The idiomatic stylistic features of the waltz and ländler in the second movement and of the march in the third are altogether divorced from their conventional trappings, unlike the ländler and march tunes that dominate many of the movements of Mahler's symphonies and the realistic dance and march episodes in *Wozzeck*. Nevertheless, they foreshadow the latter, as well as popular elements that are found in *Der Wein*, *Lulu* and the Violin Concerto. Above all, in the dance and march movements of the Three Pieces and in the passacaglia of the Altenberg songs Berg discovered the role that

150

traditional forms and traditional stylistic elements would thenceforth perform for him in restoring the possibility of coherent large-scale structure.

In May 1914, more than a year before the completion of the Three Pieces, Berg saw the Vienna première of Büchner's *Woyzeck*. He 'at once decided to set it to music', as he later wrote to Webern; and according to Gottfried Kassowitz, who was studying with him at the time, he began sketches for two scenes of the opera almost immediately. If there are such sketches, they are unknown, and it is unlikely that they ever progressed beyond the most tentative jottings. A number of passages in the opera are, however, strikingly foreshadowed in the Three Pieces: the curtain music at the conclusion of Act 1 (bars 113–19 of 'Reigen'); the tutti unisons at the conclusion of the murder scene (bars 134–5 of the 'Marsch'); the chromatically ascending transpositions of a single chord as Wozzeck drowns (bars 161–3 of the 'Marsch'). There is, also, one passage in the 'Marsch' (bars 79–84) that became a leitmotif in *Wozzeck*, the music associated with Wozzeck's portentous line in Act 1 scene ii: 'Es wandert was mit uns da unten!' ('There's something following us down there!').

The date that stands at the conclusion of the third movement of op.6, 23 August 1914, presumably refers to the completion of the scoring of that movement. The second movement was not sent off to Schoenberg until 12 months later, just about the time of Berg's induction into the army. It was not until June 1923, more than a year after Berg had completed *Wozzeck*, that the première – of the first and second movements only –

was given, under Webern's direction. In the Altenberg songs and the Three Pieces Berg had brilliantly fulfilled the prerequisites for his next project, the composition of the first full-scale atonal opera, but the latter was composed and scored without his ever having had the experience, so important to the maturing composer, of hearing an orchestral score of his in performance.

With the outbreak of the war in August 1914 Berg seems to have lost all contact for a time with the conceptual world of Büchner's drama about a poor soldier who is victimized and destroyed by the representatives of the social order. In spite of his close friendship with and enthusiasm for the writings and ideas of Kraus, who remained an intransigent pacifist throughout the war, Berg succumbed to the war fever. Along with so many other persons of learning, culture and refinement, on every side, he assumed that the material sacrifices demanded by the war would ennoble and purify society. 'The war has to continue', he wrote to Helene on New Year's eve 1914. 'If the war ended today, we should be back in the same old sordid squalor within a fortnight'. A few years later he had forgotten that he had ever supported 'the filthy war'. In a letter (27 November 1919) to Erwin Schulhoff, Berg described himself as a 'fierce antimilitarist' whose 'strongest support' in August 1914 was Kraus.

In the interim Berg had spent more than three years in the Austrian army. Though his health was precarious – seven years earlier he had suffered the first attack of the bronchial asthma which was to plague him recurrently for the rest of his life – he went through the usual basic training. After a serious physical breakdown he was reassigned to guard duty in Vienna, but the condi-

12. *Alban Berg as cadet officer, 1915*

tions of his military life remained harsh. He was eventually transferred to an office job at the War Ministry, where he remained until the end of the war in November 1918. He recalled the experience in his letter to Schulhoff:

Two and a half *years* of *daily* duty from 8 o'clock in the morning to 6 or 7 in the evening of onerous paperwork under a frightful superior (a drunken imbecile!). All these years of suffering as a *corporal*, humiliated, not a single note composed – oh, it was so horrible, that today, when I am actually freezing, and have nothing to live on, I am *happy* in comparison with those days, when life was at least *physically* endurable.

CHAPTER THREE

'Wozzeck'

In spite of the war and his illness, Berg was able to begin work on the opera once again during a period of leave in the summer of 1917, and by 19 August 1918 he had managed 'to finish something', according to his letter of that date to Webern:

It is not only the fate of this poor man, exploited and tormented by *all the world*, that touches me so closely, but also the unheard-of intensity of mood of the individual scenes. The combining of four or five scenes into *one* act through orchestral interludes tempts me also, of course. (You find something similar in the *Pelleas* of Maeterlinck–Debussy!) I have also given thought to a great variety of musical forms to correspond to the diversity in the character of the individual scenes. For example, normal operatic scenes with thematic development, then others *without* any thematic material, in the manner of *Erwartung* (understand me rightly: this is a question of *form*, not of the imitation of a style!), song forms, variations, etc.

His self-identification with the protagonist of Büchner's tragedy had made the composition of the opera a spiritual necessity. He compared himself with Wozzeck in a letter to his wife of 7 August 1918: 'I have been spending these war years just as dependent on people I hate, have been in chains, sick, captive, resigned, in fact humiliated'.

Immediately after the war Schoenberg founded the Verein für Musikalische Privataufführungen, whose aim was the presentation of 'all modern music – from that of Mahler and Strauss to the newest', under conditions that could not be obtained in the 'everyday concert world'. Admission was by subscription only, the critics

154

were excluded, no work was performed – at least that was the declared aim – until it had been allotted sufficient rehearsal time for the 'attainment of the greatest possible clarity' and 'the fulfilment of all the composer's intentions', and compositions were frequently performed twice. Berg, now a civilian, took over the practical management of the society, including organizational work, supervision of rehearsals, etc. For this he received a small monthly stipend which supplemented the income received from three private composition students and from his mother in return for his management of several houses that she owned in Vienna.

By midsummer 1919 Berg had completed Act 1 of *Wozzeck* and the first two scenes of Act 2, but a few months later a family crisis forced him to take over management of the Berghof, a responsibility which left him no time for composition and which necessitated his departure from Vienna. In April he wrote to Schulhoff:

My stay here, which is devoted to the settlement of family affairs, is coming to an end, thank God. Now I'll be able to attend once more to *myself* and *my own* affairs. After four months! *Surely* my physical ailments could all be lanced like a cyst once I were permitted to compose. But the world around me (whether it be the military, family, earning a living, etc) has robbed and continues to rob me of more than half my life.

In May the sale of the Berghof was arranged and Berg was able to return to Vienna.

For a time he considered music journalism as a means of earning a living, but his plans to assume the editorship of Universal Edition's semi-monthly magazine, *Musikblätter des Anbruch*, did not mater-

155

ialize. A letter to Schulhoff of 16 December 1920 explains his lapses as a correspondent:

During the autumn, except for a few interruptions, I was – very sick for three months. Twice in the sanitorium. A sort of nervous breakdown connected with my constant affliction of asthma. . . . Now, when things are going better with me again, I'm working as usual: giving lessons, doing musical literary work, directing Schoenberg's society during his absence (he is in Amsterdam for six months). Besides, I have just had two things of mine published (at my own expense. A pair of antique objects from the house had to pay for it): one quartet (score and parts) and four short clarinet pieces. So in addition there was the killing task of proofreading.

Meanwhile Berg was making steady progress on the opera. The short score was completed in the autumn of 1921 and the instrumentation the following spring. In the seven years since he had seen the Viennese première of the drama, war and revolution had transformed the social and political order of Europe. In the postwar world interest in the play and its author was no longer limited to a small avant-garde circle.

Büchner's drama has survived only in the preliminary drafts and sketches recovered after the author's death in 1837 at the age of 23. It was not until 38 years later that the faded and almost illegible manuscripts were deciphered, by the Galician Jewish novelist Karl Emil Franzos, and the play had to wait still another 38 years for its first theatrical production on 8 November 1913. The fact that there had been a real-life model for the protagonist of Büchner's tragedy came to light only in 1914. In 1821 a despised and poverty-stricken ex-soldier named Woyzeck – Franzos, among other errors, had misread the name of the title character – stabbed to death his faithless mistress, a crime for which he was publicly executed three years later. The story of the real

156

Woyzeck not only suggested the content of the drama but was also the source of explicit verbal motifs, employing Woyzeck's own words as reported by a court-appointed physician who concluded that Woyzeck, in spite of certain aberrations and delusions, was sane and could stand trial for his crime.

Berg's libretto is based on a second edition of the drama (Landau, 1909) which preserves Franzos's reading of the text but revises his ordering of the scenes. Aside from the omission of a number of scenes, the only material revision that Berg made in the Franzos–Landau text occurs in the scene in the doctor's study, where the doctor's anger is attributed to Wozzeck's inability to control the functions of his diaphragm rather than his bladder, a change that was indispensable if there was to be any hope of a production on the opera stage of the time. The composer's self-identification with Wozzeck during the war years explains other small revisions in this scene, like the substitution of 'beans' for 'peas', which causes the text to refer to staples of Berg's diet as a soldier in the Austrian army. The libretto in itself projects much of the formal strictness of the operatic version, even though it conforms so closely to a text that consists of a series of seemingly disconnected and fragmentary episodes. This transformation is achieved through the reduction of Franzos's 26 scenes to 15 and through the grouping of these into three acts of five scenes each, with the longer and more complex middle act bridging the symmetrically balanced outer acts. Each act is a self-contained cyclic form, and each scene a self-contained movement. Act 1, five character-pieces, is expository, each scene relating Wozzeck to a different aspect of his environment and to another person of the

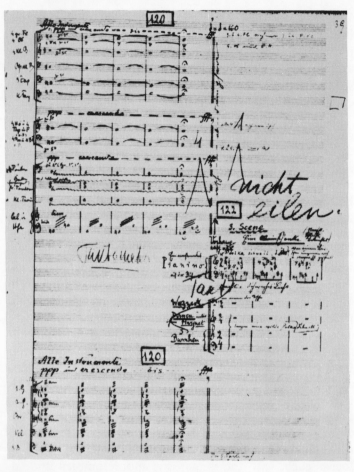

13. Autograph MS of the beginning of the third scene of 'Wozzeck', composed 1917–22

drama. Act 2 is a symphony in five movements, each scene presenting another step in Wozzeck's gradual realization of Marie's infidelity and in the gradual disintegration of the one human relationship upon which his manhood and his sanity depend. The inevitability of the dénouement, Act 3, is reflected in its five inventions, each based on a different ostinato idea.

The composer exercised an unprecedented degree of control over every aspect of the production, so that lightings, curtains, and verbal leitmotifs function structurally in the overall design. These extra-musical components are interrelated with musical cross-references to generate a highly complex yet always coherent multiplicity of associations. For example, dramatically relevant correspondences between the final scene of the opera and the first, third, and last scenes of Act 1 are established through lightings, musical leitmotifs and curtains respectively. Act 3 scene v shows a group of children, among them the now orphaned child of Wozzeck and Marie, at play. The drama proper is over and the audience is left with a crushingly nihilistic 'moral' – the essential meaninglessness of the tragedy, which is seen as nothing but a momentary distraction in an innocently callous children's world. The closing curtain and curtain music recapitulate the closing curtain and curtain music of Act 1, the point at which the fateful movement of the drama to its predestined end is initiated in Marie's seduction by the drum major. Marie's child toddles after his playmates as they go off to join the adults clustered around Marie's newly discovered corpse near the pond, and the curtain again descends upon an empty stage. The time of day, however, is morning, as it was at the beginning of the opera, so that a return to the 'normal' workaday world of the opening scene is suggested. With the exception of the curtain music, the musical material of the final scene refers to still another scene of Act 1, the cradle song episode of scene iii.

Though leitmotifs play a significant role in *Wozzeck*, they do not pervade the musical and dramatic texture as they do in Wagner's operas. Berg succeeded in his stated purpose of 'giving each scene and each accompanying interlude an unmistakable aspect, a rounded-off and finished character', in spite of a number of recurring dramatically associative motifs and even sections. (A precedent for the latter is found in the 'Liebestod' music at the conclusion of *Tristan und Isolde*.)

In relating the formal components of the musical design to the formal components of the dramatic design it was Berg's aim to achieve maximum diversity. In only three instances, for example, is the change of scene music co-extensive with what can properly be called an 'interlude' in terms of its formal function in the musical design. The second interlude of Act 1 extends beyond each of the boundaries of the change of scene, and in other instances there is no interlude at all between movements, so that the change of scene commences at some point during the movement itself. Two interludes are exactly co-extensive with curtains, one accompanying a rising, the other a falling curtain. The tempo of every curtain is precisely defined in relation to the dramatic and musical events. To cite but one example, the duration of the curtain that falls upon Marie praying for forgiveness in Act 3 scene i is seven crotchets at crotchet $= 49 (= 7 \times 7)$ and the duration of the curtain that rises upon the following scene in which she expiates

her crime through her death is six crotchets at crotchet
$= 42 (= 6 \times 7)$. The duration of the moving curtain is
thus the same in both instances, though the tempo of the
curtain music has changed.

Berg regarded it as a measure of his success that

> from the moment the curtain parts until it closes for the last time, there
> is no one in the audience who pays any attention to the various fugues,
> inventions, suites, sonata movements, variations and passacaglias – no
> one who heeds anything but the social problems of this opera which by
> far transcend the personal destiny of Wozzeck.

In fact, it is above all through the autonomy of the
design and the order that governs every component that
a characteristic theme of Büchner's is all the more powerfully projected, a theme discussed by A. H. J. Knight
in connection with Büchner's prose fragment, *Lenz*:
'The perpetual, poignant, horrifying contrast between
men, so hopelessly involved in a universe not made by
them and hostile to their interests and efforts, and that
universe itself, as it mechanically functions in the mani-
festations of external nature'.

The occasional presence of diatonic tonality in
Wozzeck has been frequently noted, but of more interest
and significance is the establishment of specific pitches
or collections of pitches as focal elements in non-
diatonic contexts. The formal components of the sonata-
allegro movement (Act 2 scene i) are effectively defined
and characterized by means of referential harmonic
units, so that there is a valid analogy with the means
used to define and characterize such components in
traditional tonality. Devices that are found in earlier
atonal music are given large-scale structural implications for the first time in *Wozzeck*. Chromatic inflection,
for example, generates a chord progression that is the

161

basis of a complete scene and its adjoining interludes (ex.9). Very soon after the completion of *Wozzeck*, 12-note serialism was conceived as a comprehensive and general basis for the organization of atonal music, but a close study of the musical language of *Wozzeck* suggests

Ex.9

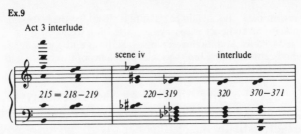

Act 3 interlude

that other lines of development were viable possibilities as well.

Since he was still without a publisher when he completed the work, Berg decided to publish the vocal score of *Wozzeck* himself, as he had each of his earlier works that had so far appeared in print. Alma Mahler, the composer's widow and a close friend of the Bergs, raised the money for this venture, and in January 1923 Berg sent out printed cards announcing the publication. In April 1923 Viebig's perceptive study of the new work appeared, an enthusiastic appraisal by a writer who was personally unknown to Berg. With the acceptance of the work by Schoenberg's publisher, Universal Edition, Berg was encouraged to hope for an early production. In the summer of 1923, at the first annual festival of the ISCM in Salzburg, where Berg was present for the performance of his quartet, Scherchen expressed his interest in conducting excerpts in a concert version during the following season. The concert suite, *Drei*

Bruchstücke aus 'Wozzeck', was introduced in Frankfurt in June 1924 and was an immediate success.

Even before this, Kleiber, the young music director of the Berlin Staatsoper, had given Berg assurances of his intention to produce the opera. The project soon became a subject of partisan controversy involving conflicting ideological, political, professional and aesthetic positions. There had been opposition to Kleiber's appointment and as the date of the première, 14 December 1925, approached, his critics, describing Berg as 'the most intransigent of all the Schoenbergians', made *Wozzeck* an issue in the campaign against Kleiber. Right-wing papers drew parallels between music and politics: 'Where anarchism in political life will take the nations may be a question of the future for politicians; where it has taken us in art is already manifest. The young talents have had their fling and left us a rubbish dump on which for years henceforth nothing will grow or prosper'. There were warnings of the imminent economic collapse of the state theatres and false reports of riots at the opera house. But on the whole the public and the critics sensed, with Adolf Aber, chief critic of a Leipzig paper, that *Wozzeck* was a work 'whose great intrinsic qualities will still be valid when there will be no one left who remembers the current crisis of the Berlin opera'. There could be no question of the extraordinary success of the opera and in the same season the Berlin opera followed the première with nine more performances.

CHAPTER FOUR
The 'Lyric Suite' period

In the two years between his private publication of the vocal score of *Wozzeck* and its première in Berlin, Berg completed a large work of a very different character. The Chamber Concerto for piano, violin and 13 wind instruments represents, as do Schoenberg's Serenade and Piano Suite and Webern's String Trio, a definitive departure from the expressionist ethos of 'free' atonality. The turn towards a more 'objective' style which revived the clear sectional divisions and literal repetitions of the Classical forms was consistent with Schoenberg's aim in formulating the 12-note system: '[to lay] the foundations for a new procedure in musical construction which [seems] fitted to replace those structural differentiations provided formerly by tonal harmonies'. But it is also consistent with the general trend of the period, which saw the rise of 'neo-classicism' as a movement opposed to the direction taken by Schoenberg and his followers. The Chamber Concerto shows the influence of Schoenberg's new technical procedures in its frequent use of 12-note collections and in its dependence on the basic transformations – prime, inversion, retrograde inversion and retrograde – of the 12-note system. But it is still far from that system, since these collections are merely occasional special formations that do not determine the totality of pitch-class relations of the context in which they occur, and the basic transformations are employed only as formal devices rather than as the complementary aspects of an exclusive referential ordering of pitch classes. Each variation of the first movement, 'Thema scherzoso con variazione', unfolds a single transformation of the 30-bar theme in the following sequence: prime, retrograde, inversion, retrograde inversion, prime. In the second movement, Adagio, a large *ABA* design is repeated in retrograde and, since the two *A* components are originally related as prime to inversion, the second half of the movement unfolds the remaining basic transformations, retrograde inversion and retrograde. The third movement comprises six sections, each of which simultaneously paraphrases the two formal divisions that occur in corresponding positions in the two preceding movements.

Despite its many attractive details, the Chamber Concerto is the least successful of Berg's mature works. The formal design seems schematic and contrived in its dependence on paraphrase procedures that rigorously reproduce the same relative formal dimensions, and the harmonic language seems static and undifferentiated. Berg's reference, in his open dedicatory letter to Schoenberg, to 'passages that correspond to the laws set up by you for "composition with twelve notes related only to one another" ' suggests that he understood nothing at all, at the time, of these 'laws'. The work nevertheless shows Berg, like Schoenberg, searching for objective principles of general relevance to atonal composition, and it is a first large step in the evolution of the musical language of his masterpiece, *Lulu*.

In a letter to Webern dated 12 October 1925, Berg referred to his 'first attempt at strict 12-note serial composition', a new setting of the poem by Storm,

Schliesse mir die Augen beide, which he had provided with a tonal setting in 1907. The set is the same symmetrical all-interval series (ex.10*a*) on which he based the first movement of his next work, the Lyric Suite for string quartet. The overall concept of the latter, 'six rather short movements of a lyrical rather than symphonic character', is mentioned in the same letter to Webern, though Berg was still in the very first stages of its composition.

Ex.10

Berg explained in the analytical notes he prepared for Rudolf Kolisch, whose quartet gave the first performance of the work on 8 January 1927, that the numerous thematic connections among the different movements are not 'mechanical', but motivated rather by 'the large unfolding (the continuing intensification of mood) within the *whole* composition'. This 'intensification' is realized in the macro-structure through the increasingly diverging tempos of the six movements: Allegretto giovale, Andante amoroso, Allegro misterioso, Adagio appassionato, Presto delirando, Largo desolato. The successive psychological states that are denoted in these tempo designations are successive stages in what Adorno called a 'latent opera', and Berg himself explained progressive changes in the structure of the original 12-note series in the course of the work as implying 'submission to fate'.

The only consistently 12-note movements are the first, third (exclusive of the trio) and sixth, but 12-note configurations are included among the principal themes of the intervening non-serial movements, and there are 12-note episodes as well, in the second and fifth movements, that anticipate the 12-note movement which follows each. In addition to these serial connections between the different movements there are thematic and even sectional quotations from movement to movement. Yet such interrelations do not diminish in any way the autonomy of the individual movements. As in *Wozzeck*, this very autonomy points up the referential character of shared thematic material and quotations, which ultimately seem to acquire leitmotivic implications that justify Adorno's characterization of the work. That these themes and quotations are indeed leitmotifs was confirmed in January 1977, when a copy of the miniature score profusely and carefully annotated by the composer was discovered (see Perle, 'The Secret Programme', 1977). The annotations unfold a secret programme inspired by Berg's love for Hanna Fuchs-Robettin, the wife of a Prague industrialist and sister of Franz Werfel. Her initials combine with Berg's to give the basic cell of the work, B–F–A–B♭ (in German notation, H–F–A–B).

Those special features that distinguish Berg's 12-note practice in general from that of Schoenberg and Webern are all found in the Lyric Suite: first, dodecaphonic (i.e. consistently based on a 12-note set) and non-dodecaphonic episodes occur in the same work, not only from movement to movement, but also sometimes within the same movement. Secondly, different sets are employed, even within a single movement. These include tropes (i.e. 12-note sets partitioned into two or more segments and defined by the unordered pitch-class content of these segments) as well as the Schoenbergian type of ordered sets (i.e. series). Thirdly, the series are employed in various cyclical permutations. Fourthly, an explicit linear contour tends to be consistently associated with a given series. Finally, since the referential contour remains easily recognizable under inversion, but not under the other basic transformations of Schoenberg's 12-note system, retrograde and retrograde inversion, the latter do not occur as independent forms of the series. The first movement of the Lyric Suite is based on a symmetrical series (ex.10*a*) whose R and RI forms are respectively identical with the tritone transpositions of P and I. The R and RI forms of the 12-note series of the third movement and the two 12-note series of the sixth movement never appear independently of palindromes that totally reverse antecedent passages restricted to P and I forms.

It is a mistake to regard these distinctive features of Berg's 12-note practice as 'licences' that show an indifferent attitude to the systematic aspects of 12-note composition. The implications of set structure are as consistently realized as in the music of Schoenberg and Webern, but in different ways. If, for example, Berg chose to employ three different sets in the first movement of the Lyric Suite, this was not because of a failure to appreciate that concept of the set as a unifying

168

device which led Schoenberg to observe that it did not seem 'right' to him to use more than one series in a piece. Defined in terms of their hexachordal content, all three sets turn out to be representations of a single trope (ex.10), and their various transpositions and cyclical permutations are used to establish a hierarchy of relationships that is a remarkably appropriate basis for the sonata-allegro design of the movement.

The Kolisch Quartet achieved an outstanding success in its frequent performances of the Lyric Suite, which, for all its subjective and tragic character, remains one of the most brilliant and effective virtuoso display pieces in its genre. When it was performed before an international audience for the first time, at the Baden-Baden Festival on 16 July 1927, the work had to be repeated. Among the auditors was Bartók, who performed his Piano Sonata on the same occasion. Bartók's Third Quartet is dated September 1927, so it is not impossible that some of the novel features in the string writing of that work were influenced by Berg's unprecedented exploitation of the technical and sonic resources of the medium.

Wozzeck was meanwhile proving its staying power with a second season of performances in Berlin and new productions in other cities. In Prague, as in Berlin in the previous season, there was a politically motivated campaign against the opera, the opposition in this instance being led by Czech nationalists who demonstrated their anti-German sentiment by attacking the same work (in spite of its translation into Czech) that the German nationalists had attacked as violating the spirit of German art. Disorder during the third performance caused the police to clear the theatre and to place a ban on further performances. A third production, in

Leningrad in June 1927, was, according to Berg's telegram to his wife, 'a huge, tumultuous success'. A successful production in Oldenburg on 5 March 1929 proved that the work was not beyond the means of a provincial theatre, and in the following season alone there were 40 performances in eight German opera houses. Characteristically, the composer's own city, Vienna, did not produce the opera until 30 March 1930, only one year before the American première under Stokowski.

By September 1928, two months before his meeting with Wedekind's widow to conclude financial arrangements for the rights to the drama, Berg had already composed more than 300 bars of his second opera, *Lulu*. A commission for a concert aria from the Viennese soprano Ružena Herlinger in the spring of 1929 led Berg to interrupt work on the opera, and by the summer's end he had completed *Der Wein*. Redlich has pointed to the significance of the aria as a study for the opera, as shown in its vocal style, in the choice of text (Baudelaire 'wrote these poems in a spirit of revolt, aimed at the complacent philistinism of his epoch, and anticipating Wedekind's later indictment of the bourgeoisie of the "fin de siècle" '), in its formal design, and in its 'colour and scoring'. Piano and saxophone are used as they are in *Lulu* and as they already had been in Schoenberg's *Von heute auf morgen*, both as an integral part of the orchestra and in reference to the idiom of commercial popular music. Both *Lulu* and *Der Wein* exemplify Berg's concept of the proper role of the voice in opera as set forth in an essay that he published in the same year in which he composed the aria. He emphasized the primacy of the singer and the importance of bel canto, assigning to opera the task, 'above all, of serving the human voice and promoting its rights – which, indeed, it had almost lost in the musico-dramatic works of recent decades in which operatic music, as Schoenberg has said, has often represented nothing else than a "symphony for large orchestra with vocal accompaniment" '.

In its use of a set that may be partitioned into pitch collections equivalent in form to harmonic categories of the diatonic tonal system, *Der Wein* anticipates the Violin Concerto. The series in its principal set form unfolds what in other contexts would be a D minor diatonic hexachord, a G♭ major triad, and a V⁷ of D♭ major (ex.11). The initial entry of the voice stresses the implication of D minor through a melodic cadence on

Ex.11

Ex.12

Des Wei-nes Geist   be-gann im Fass zu sin - gen

the fifth note of the series (ex.12). The pervasive harmonic texture suggests an enormously extended post-Wagnerian chromatic idiom that cannot in any real sense be said to be derived from the set. Both *Der Wein* and *Lulu* depart significantly in this respect from the 12-note practice of Schoenberg and Webern. The basic series of *Der Wein* is freely and continuously transformed through cyclical permutations and revisions in ordering and serves above all as a common source of melodic and thematic elements.

171

Der Wein is in its formal plan a combination of sonata-allegro and ternary form, with the three poems which make up its text respectively set as sonata exposition, contrasting middle section in the style of a scherzo, and recapitulation and coda. This intermixing of traditional formal types is resumed on a much larger scale in *Lulu*.

CHAPTER FIVE

'Lulu'

The new direction in Berg's evolution that was to culminate in the music of *Lulu* becomes suddenly apparent in his work immediately following the completion of *Wozzeck*. His involvement with its dramatic and literary content began much earlier. *Wozzeck* was the result of a special experience – the impact of Büchner's still-unknown play and of the years of war – but *Lulu* is a work that Berg was fated to write, a work that is foreshadowed in the literary interests, social concepts and character-forming experiences of his adolescence and youth. An American friend of his youth, Frida Semler, who spent the summers of 1903 and 1904 at the Bergs' country home while her father attended to his business affairs in Vienna, described their summer reading: 'In the evening we read Ibsen aloud with divided parts. In 1903 Ibsen was a discovery, and we thrilled over Baumeister Sollness and Hedda and *Ghosts*. It was in the second summer that we all read – not aloud – Schnitzler's *Reigen* and Wedekind's *Erdgeist*'. In the following spring Kraus staged the sequel to *Erdgeist*, *Die Büchse der Pandora*, which was still under the censor's ban in Germany. Wedekind himself, playing the part of Jack the Ripper, met his future wife in the person of the young actress who was playing Lulu. Tilly Wedekind recalled the occasion in her memoirs:

In the hall, filled to capacity, there sat, one among many, a young man of 20 who looked like an angel. Decades later the world became aware of

14. *Alban Berg (left) with Anton Webern*

the lasting impression that the play, the production, and the [introductory] talk by Karl Kraus had made on him. His name was Alban Berg.

18 months later Berg wrote to Frida Semler:

Wedekind – the really new direction – the emphasis on the sensual in modern works!! ... At last we have come to the realization that sensuality is not a weakness, does not mean a surrender to one's own will. Rather is it an immense strength that lies in us – the pivot of all being and thinking.

To appreciate the significance of these words one must know the social context in which they were uttered, the 'sticky, perfumed, sultry, unhealthy atmosphere' which Zweig described in *The World of Yesterday*, the 'dishonest and unpsychological morality of secrecy and hiding [that] hung over us like a nightmare'. Berg's letters to his fiancée occasionally give a glimpse of real-life situations that suggest a more direct identification with the content of the Lulu plays. Unable to keep an appointment with Helene, he explained that it was because his lesbian sister, Smaragda, 'last night tried to poison herself with gas. Apparently it didn't do her much harm physically, but mentally she's in complete despair, poor soul'. In another letter he mentioned 'a prostitute, my sister's present friend', and echoing Kraus's criticism of the sham morality of Viennese society added that he found 'a prostitute's position no more or less offensive than associating with people whom you and many others consider quite unobjectionable'.

For some time after he had begun work on the composition itself, Berg continued to occupy himself with the formidable problem of transforming Wedekind's two Lulu plays into a libretto. The lengthy and complicated text had to be drastically condensed and simultaneously clarified, while maintaining, as Berg explained in a letter to Schoenberg, 'Wedekind's idiomatic characteristics in the process'. The *dramatis personae* had to be converted into personalities that could be effectively characterized by musical means, and Wedekind's dramatic conception amended and adapted in such a way as to justify this conversion. Berg not only solved these problems, but in doing so he created a work that is more coherent, more complex though less complicated, more profound, and in every way more effective dramatically than the original.

Only the five principal characters – Lulu, Schigolch, Dr Schön, Alwa and Countess Geschwitz – retain the names given them by Wedekind, the others being identified merely by their calling or title (the Painter, the Acrobat, the Schoolboy, the Marquis, etc.). Berg thus sharpened a distinction that Wedekind himself had made between the identity of Lulu, each of whose three husbands has his own name for her, and that of the other characters. The seven acts of the two plays are converted into the seven scenes of the opera according to the following plan:

Erdgeist	*Lulu*, part 1
	Act 1
Act 1	scene i
Act 2	scene ii
Act 3	scene iii
	Act 2
Act 4	scene i
Die Büchse der Pandora	*Lulu*, part 2
Act 1	scene ii
	Act 3
Act 2	scene i
Act 3	scene ii

176

Part 1 shows Lulu in her ascendant phase, culminating in her marriage to Dr Schön, the newspaper publisher and powerful man of affairs whose mistress she
has been for many years, and concludes with the murder
of Dr Schön. This he brings upon himself when he
discovers Lulu with his son, Alwa, and, in a mad climax
to his earlier attempts to sever his ties to her by marry-
ing her off to someone else, hands her his revolver and
demands that she shoot herself. Part 2 shows Lulu in
her descendant phase. She returns, after her escape from
prison, to the murdered man's apartment to meet Alwa,
who has helped to plot her escape. In the final act they
take refuge in a gambling salon in Paris, but are forced
to run away a second time when the Marquis threatens
to turn Lulu in to the police if she persists in her refusal
to be sold into white slavery. She ends as a common
streetwalker in London, dying at the hands of Jack the
Ripper. The principal formal break in the opera occurs
between the two scenes of Act 2 and is marked by a
musical interlude intended to accompany a three-minute
silent film representing the action implied between the
conclusion of the first play and the beginning of the
second: Lulu's arrest, commitment for trial, trial, im-
prisonment, and then – visually analogous events in
reverse order accompanied by a retrograde version of
the film music – Lulu's removal from prison because of
illness, her physical examination, her commitment to
hospital, and her escape.

Three different concepts of operatic form are joined
by Berg. The largest dimensions of the work depend on
recapitulative episodes which become increasingly extensive until, in the final scene, they dominate the material completely. The principal key to these is found in the

177

composer's assignment of multiple roles to individual performers. Each of Lulu's three victims in the first half of the opera – the Medical Specialist, the Painter, Dr Schön – is respectively paired with one of the three clients, symbolic avengers of those who have lost their lives because of their love for her, whom she brings in from the street in the final scene – the Professor, the Negro and Jack the Ripper. The telescoped recapitulation, in the last scene, of the music of the first half of the opera corresponds to these doublings. Another level of formal integration in *Lulu* depends on the use of large-scale 'absolute' forms. Each act is dominated by a single formal structure whose subdivisions are distributed throughout the act and separated by independent formal units. This dominant formal component of each act is associated with the principal dramatic idea of the act. Thus the Sonata-allegro movement of Act 1 represents Dr Schön's vain struggle to break with his mistress, the Rondo of Act 2 represents Alwa's passionate attachment to Lulu, and the Theme and Variations of Act 3 represents the nadir of Lulu's career, when she has descended to the life of a streetwalker. At a more immediate level, the formal design of *Lulu* returns to the classical 'number' opera in its use of highly individualized and self-contained small forms. Berg called attention to these in the score by the use of such titles as Recitativ, Canzonetta, Arietta, Lied, Duett, Arioso, Kavatine, Interludium, etc.

The doubling of the roles of Lulu's victims in part 1 with those of her clients in the final scene is essential to the dramatic structure of the opera. Other doublings are merely a matter of convenience and economy. In specifying the latter as well, and pointing them up by musical means (leitmotifs, serial procedures, vocal range and style, formal recapitulations), Berg brought that which lies outside the drama, the purely practical aspects of the production, into the work itself. This is consistent with the change that he made in Alwa's profession: in Wedekind's version he is a poet and playwright who identifies himself as the author of *Erdgeist*; in the opera he is a composer whose real identity is revealed in the orchestra when the opening chords of *Wozzeck* are heard as Alwa says of Lulu 'One could write an interesting opera about her'.

Ex.13 shows the primary P and I set forms of the basic series. The partitioning of the two set forms shows

Ex.13

their common hexachordal pitch-class content. (The transposition numbers, 0 and 9, are the respective pitch-class numbers, 0 for C and 9 for A (the 9th semitone above C), of the initial element of each set form.) A hierarchy of differentiated harmonic areas is established through the common segmental pitch-class content of selected forms of various sets. For example, Dr Schön's series at I_9 and that of his son Alwa at P_4 are identical as to both trichordal and hexachordal pitch-class content (exx.14–15). The same series are each maximally invariant as to hexachordal content with the basic series at P_0 and I_9, in that corresponding hexachords have in

179

common the largest number of pitch classes (five) that their respective structures will permit.

A new type of set that plays an important role in the opera is the 'serial trope', i.e. a set partitioned into internally ordered segments that are individually subjected to such serial operations as will not revise segmental pitch-class content. Three variant forms of Schigolch's serial trope are illustrated in ex.16. The

chromatic scale is the ultimate source of all the variants of Schigolch's serial trope. The association of this set with Lulu's shadowy background is consistent with the relation between Schigolch and Lulu. One first takes the old beggar to be Lulu's father, but later it appears that he is really a former lover, a relic of some vague past

long before the beginning of the present drama. Since he
alone seems to know who she really is, and since he
continues to take a proprietary interest in her until the
very end, he may still be regarded as her 'father' in a
figurative sense. At the end only the frail, asthmatic old

Ex.17

man survives, unchanged, having left Lulu's garret for a
drink at the pub just before her return from the street
with Jack.

Long-range as well as local harmonic associations are
established through the sharing of pitch-class collections
among different set forms. An example is the relation
between Dr Schön's Arietta ('So this is the evening of
my life') near the beginning of Act 2 scene i and Lulu's
Arietta ('You can't hand me over to the police!') ad-
dressed to the murdered man's son at the conclusion of
the same scene. These are musical as well as dramatic
counterparts. The common hexachordal content of Dr
Schön's series at I_9 and Alwa's series at P_4 associates the

181

two numbers harmonically. The characterization of every series in the opera by a referential contour provides another means: shared melodic cells, by which connections are established among different series. At the rise of the curtain on the first scene of Act 1, for example, Alwa (the composer himself) asks, 'May I come in?' The words are set to a segment of the basic series that coincides with a segment of Alwa's series, though there is no statement of the latter at this point (ex.17).

These and any other examples from the opera that might be quoted reaffirm a point already made in the discussion of the Lyric Suite: Berg's 12-note music is not less systematic than that of Schoenberg and Webern, as has been generally assumed. If anything, it is even more systematic, but its methods and implied axioms are very different. The composer himself camouflaged (to himself as well as others) his divergence from Schoenberg's concept of 12-note composition by promulgating, through his authorized biographer, specious analyses of his music. This is the source of the far-fetched arguments according to which all the sets in *Lulu* are supposed to be derived from the basic series. Indeed, one of the two basic cells of the opera, a four-note collection that occurs throughout the work independently of any 12-note set, is also the generator of a most important 12-note trope (ex.18) that represents the staged world of the drama and the fatality and enchantment of its central character in a general sense. The second of these basic cells (ex.19) is similarly employed as an independent structure, but it also occurs as one of the segments of Countess Geschwitz's trope (ex.20) and, in a special ordering, as the initial five-note segment of the I_9 form of Dr Schön's series (ex.14).

182

Ex.18

Ex.19

Ex.20

Two general concepts of post-diatonic pitch organization that play an important role in all of Berg's work beginning with the Quartet op.3 – inversional symmetry and the unfolding of interval cycles – are greatly enlarged in their significance and pervasiveness in *Lulu*. A summary illustration of the first of these is found in the finale of Act 2 scene i, bars 362–561, where a variety of compositional elements – Dr Schön's series, Countess Geschwitz's trope, the basic series, the basic cells, arrays of perfect-4th chords – unfold one or the other of the two sets of inversional relations represented in exx.21 and 22. The thematic employment of whole-tone, semitone and perfect-4th cycles was illustrated in ex.2. A deeper significance is assigned to an interval cycle in connection with the basic series of *Lulu*. The primary P and I set forms (ex.13) are each a special ordering of the correspondingly partitioned cycle of 5ths (ex.10*b*). Transposition by a perfect 5th revises the pitch-class content of each hexachord through the replacement of a single element by its tritone. A hierarchy of relations is thus generated between transpositionally related set forms, with maximum intersection (five elements in common) of hexachordal content between set forms separated by a 5th and total exchange of hexachordal content between set forms separated by a tritone. Other interval cycles partition the total content of the semitonal scale into mutually exclusive groups. The whole-tone cycle, for instance, partitions it into two six-note collections, each of which remains invariant as to content under transposition by any even interval. It is principles of this sort which provide a basis for musical coherence in *Lulu*, and not its supposed derivation from one primordial note row.

Ex.21

Ex.22

In its organization of rhythm and tempo *Lulu* is a similarly advanced and summational development of concepts introduced in Berg's earlier works, *Wozzeck*, the Chamber Concerto and the Lyric Suite, but it is difficult to think of precedents or parallels in the work of any other composer. In its overall dimensions the whole last scene unfolds a single large-scale, terraced *ritardando*, through successive changes in the principal

184

tempos. The reciprocal relation between tempo and rhythm becomes a controlled compositional device, as it already had in *Wozzeck*, a means of modulating from one tempo to another and of establishing both local and large-scale connections. There is a special rhythmic theme, the *Hauptrhythmus* (ex.23), and its various

Ex.23

derived forms, a concept that has been traced back to the Altenberg songs and that is the basis of a whole scene, the Invention on a Rhythm (Act 3 scene iii) in *Wozzeck*. In *Lulu* this 'fate rhythm' is assigned a structural and leitmotivic role that pervades the entire work.

CHAPTER SIX
Last years

The short score of *Lulu* was finished in the spring of 1934. Berg was again in straitened circumstances. The success of *Wozzeck*, which by 1932 had been staged in 17 German cities, had greatly improved his economic situation for a time, but with the growing political power of the Nazis the work was increasingly excluded from the repertory, in spite of the composer's 'aryan' credentials. At the end of February 1933, four weeks after Hitler's appointment as chancellor, Berg was in Munich as a member of the programme jury of the Allgemeine Deutsche Musikverein. It was carnival time. 'The whole town and all its inhabitants are quite drowned in carnival din, masks and confetti', Berg wrote to his wife. 'And on top of that the news of the Reichstag fire. Dancing on a volcano!' The sessions with the jury were 'strenuous'. 'The Nazis have to be considered so much that Schoenberg, for instance, drops out, also non-German names like Pisk and Jelinek, who in different circumstances would certainly have been chosen.' In March the dismissal of Jewish musicians from civic posts began. Among these was Schoenberg, who had held the post of professor at the Prussian Academy of Arts in Berlin since 1925. In a letter to Webern (29 June 1933) Berg wrote 'My utter depression over *these* times has for a long time now impaired my ability to work', and a few days later he wrote of his concern for Schoenberg, who had left for Paris: 'Now at

186

the age of nearly 60, expelled from the country where he could speak his mother tongue, homeless and uncertain *where*, and on *what*, to live'.

It had been Berg's hope that Kleiber would conduct the première of his new opera, but a production in Germany was now out of the question. Berg was nevertheless counting on Kleiber for a first performance of the projected suite of pieces from *Lulu* which he planned to score first of all. 'I think that we could risk doing it even in Germany', he wrote to Kleiber. The composer was not present for the première of the *Symphonische Stücke aus der Oper 'Lulu'*, which took place on 30 November 1934. Four days later Kleiber resigned his post.

In January 1935 Berg received a commission for a concerto from the American violinist Louis Krasner. The death of Alma Mahler's 18-year-old daughter, Manon Gropius, on 22 April led Berg to interrupt his work on the instrumentation of *Lulu* in order to compose the concerto as a memorial for the dead girl. Working at what was for him a remarkable pace, Berg completed the short score of the Violin Concerto in mid-July and the full score on 11 August. As always with Berg, the 'programme' of the work unfolds within a large-scale design of strict formal proportions. The four movements are grouped into two pairs, with a pause only between the second and third movements. The opening Andante, with its clear subdivisions into introduction, principal subject, subordinate subject, concluding subject and codetta, and the following scherzo (Allegretto) are 'classical' in the symmetry and balance of their phrase structure. Fast and slow tempos are reversed and intensified in the second pair of movements, an Allegro in the free style of a cadenza and a concluding Adagio based on the Bach chorale *Es ist genug*. The first part offers us a musical 'portrait' of the girl, the second a representation of catastrophe and, finally, submission to death, and transfiguration.

The basic set in its prime form and principal pitch level is shown in ex.24. A direct statement of the set itself serves as the primary melodic component of the principal subject of the first movement. The whole-tone segment which concludes the series coincides with the initial motif of the chorale: Alternate notes of the series of 3rds that precede this segment are the source of another basic leitmotif, the arpeggiated open-string figure of the soloist's first entry. The harmonic material of the traditional tonal system is suggested in the 5th-related major and minor triads of the set. A consistent context is thus provided for the interpolated tonal quotations – a Carinthian folksong in the coda of the scherzo and Bach's own harmonization of the chorale melody in the Adagio. Some episodes in the Violin Concerto are both tonal and serial, some are tonal but not serial, some are serial but not tonal, and still others are neither tonal nor consistently serial. The many intersections among different set forms usually make an unambiguous serial analysis impossible, even where the texture is more or less consistently dodecaphonic. There is total intersection between P and RI (and therefore I and R) set forms transpositionally related as in exx.24 and 25 and cyclically permuted relative to each other. Thus R and RI are again eliminated as independent transformations.

Ex.25

In mid-August, soon after completing the orchestral score of the Violin Concerto at the Waldhaus, Berg suffered a painful abscess on his back, presumably incurred through an insect sting. In spite of treatment, the infection persisted. Though he was in pain for much of the time, he remained at the Waldhaus until mid-November, continuing to work on the score of *Lulu*. On 11 December the *Lulu* suite was played in Vienna for the first time, and Berg, who had not yet heard the work, was in the audience, though gravely ill. He was admitted to hospital on the 17th and died of general septicaemia one week later, on the morning of 24 December 1935.

The Violin Concerto was introduced posthumously on 19 April 1936, at an ISCM Festival in Barcelona. Thus what Berg had presented to the world as a memorial to another became the composer's own requiem. What the world did not know, and what has only recently been revealed (see Jarman, 1982), was that Berg had planned the concerto as a double requiem – that he had taken advantage of the inherently ambiguous character of programmatic expression in music to conceal an alternative and equally authentic programmatic conception beneath the one that he had offered to the public. A first clue is given in the bar count of the Introduction, which the composer explicitly indicates in the score: 'Introduction (10 Takte)'. From Berg's own annotations in Hanna Fuchs-Robettin's copy of the Lyric Suite we know that 'our numbers', Berg's and Hanna's, were respectively 23 and 10. A second clue is given in the curious expression markings that invariably accompany each phrase of the chorale, above all the 'amoroso' assigned in the score to every statement of the four-note closing figure, a marking that is hardly appropriate to the text and implication of the original tune. A third clue is found in the original text, of which there is no hint in the score, of the Carinthian folksong, which refers to the singer's liaison with one 'Mizzi'. 'Mizzi' was also the nickname of a servant girl in the Berg family household, whose intimacy with the young Alban resulted in the birth of an illegitimate daughter. These clues point to a whole system of cryptographic bar-counts, metronome marks and musical ciphers whose meaning we can deduce from the annotated score of the Lyric Suite and from what we know of Berg's interest in the numerological theories of Wilhelm Fliess, a Berlin biologist and an early mentor of Sigmund Freud. In the alternative programme of the Violin Concerto the two statements of the folksong, in the Scherzo of the first part and the closing Adagio of the second, respectively represent Berg's first consequential love affair, with the servant girl who was the mother of his child, and his last, idealized relationship with Hanna Fuchs-Robettin. It is the composer's own mortality that is represented in the catastrophic violence of the Allegro and the resignation of the Adagio. The final reminiscence of the *Ländlermelodie* and the coda are a musical paraphrase of feelings repeatedly expressed in his letters to Hanna: 'No one can take from me the certainty of our union in a later life', he wrote in May 1930, and 'How many more years – until eternity, which belongs to us???' on 9 December 1931.

Moved as we are by this work and the circumstances that motivated its composition, it is to be regretted that it prevented Berg from completing the orchestration of an infinitely more important work, one of the supreme masterpieces in all of operatic literature. The full score of the last act of *Lulu* was already well advanced when death stayed his hand. The interlude between scenes i and ii and portions of the conclusion of scene ii had been scored as part of the *Lulu* suite, and when Berg returned to the opera after completing the Violin Concerto he was able to continue the full score as far as the first 268 bars of Act 3 scene i. In his article on *Lulu* in the October 1936 issue of the *Musical Quarterly* Willi Reich wrote:

Berg left a complete and very carefully worked out preliminary score of *Lulu*. Only the instrumentation of a few places in the middle of the last act was not finished and this could easily be carried out from the given material by some friend familiar with Berg's work.

Erwin Stein's vocal score of Acts 1 and 2 was published in the same year, with a prefatory note by the publisher stating that the vocal score of Act 3 would be published at a later time. Stein's reduction of Act 3 was, in fact, completed, on the basis of Berg's short score, but its publication was interrupted after 70 pages had already been engraved.

In Austria the Nazi takeover was imminent and there was no longer a German or Austrian opera house where *Lulu* might be staged. On 2 June 1937 the opera was performed for the first time, in Zurich. Of the music for Act 3, only those portions that Berg had incorporated in the *Lulu* suite were presented, as 'background' music to an adaptation of the final episode of the drama, the murder and death of Lulu and the Countess. When the opera was revived after the war Berg's widow refused to release the unpublished material of the third act. The opera continued to be performed with a makeshift finale along the lines of the Zurich première, a finale which retrospectively falsified the dramatic and musical content of the two preceding acts and destroyed the symmetry of the work. Helene Berg's death, on 30 August 1976, gave rise to hopes that the publication of Stein's vocal score of Act 3, announced in 1936, would at last be achieved. Stein's reduction is a faithful and expert representation of the composer's short score; it was finally published in 1979, in an edition by the Viennese composer, Friedrich Cerha, who had, with the permission of the publisher, secretly completed the full score long before, while the widow was still alive. The first complete performance of the opera was given at the Paris Opéra on 24 February 1979.

Numbers in the right-hand column denote references in the text.

WORKS

op.

— Many songs, 1900–05; for details see Chadwick (1971)

— Seven Early Songs, 1v, pf, 1905–8, rev. and orchd 1928; orch version perf. Vienna, 6 Nov 1928; pf version (1928), orch version (1969): Nacht (C. Hauptmann), Schilflied (N. Lenau), Die Nachtigall (T. Storm), Traumgekrönt (R. M. Rilke), Im Zimmer (J. Schlaf), Liebesode (O. E. Hartleben), Sommertage (P. Hohenberg) — 138

— Schliesse mir die Augen beide (Storm), 1v, pf, 1st setting, 1907; pubd in *Die Musik*, xxii (1930); repr. (1955) — 166

— An Leukon (J. Gleim), 1v, pf, 1908; pubd in Reich (1937)

— Variations on an Original Theme, pf, 1908; pubd in Redlich (1957)

1   Piano Sonata, ?1907–8; perf. Vienna, 24 April 1911; (1910) — 138, 141, 149

2   Four Songs, 1v, pf, ?1909–10 (1910): Schlafen, schlafen (C. Hebbel), Schlafend trägt man mich (A. Mombert), Nun ich der Riesen Stärksten (Mombert), Warm die Lüfte (Mombert) — 140

3   String Quartet, 1910; perf. Vienna, 24 April 1911; (1920) — 140, 143, 145, 149, 163, 183

4   Fünf Orchesterlieder nach Ansichtskartentexten von Peter Altenberg, 1v, orch, 1912; 2 nos. cond. Schoenberg, Vienna, 31 March 1913; all 5 cond. Horenstein, Rome, 1952; vocal score (1953), full score (1966): Seele, wie bist du schöner, Sahst du nach dem Gewitterregen, Über die Grenzen des All, Nichts ist gekommen, Hier ist Friede — 144, 145, 146, 147, 149, 150, 151, 152, 185

5   Four Pieces, cl, pf, 1913; perf. Vienna, 17 Oct 1919; (1920) — 144, 145, 147, 148, 149, 150

6   Three Pieces, orch, 1914–15; nos.1–2 cond. Webern, Berlin, 5 June 1923; all 3 cond. J. Schüler, Oldenburg, 14 April 1930; (1923): Präludium, Reigen, Marsch — 148, 149, 150, 151, 152

7   Wozzeck (opera, 3, Büchner), 1917–22; cond. Kleiber, Berlin, Staatsoper, 14 Dec 1925; (1923) — 141, 145, 146, 150, 151, 154ff, 158, 164, 167, 169, 173, 179, 184, 185, 186

— Drei Bruchstücke aus 'Wozzeck', S, orch; cond. Scherchen, Frankfurt, 11 June 1924; (1924) — 164

— Chamber Concerto, pf, vn, 13 wind insts, 1923–5; cond. Scherchen, Berlin, 27 March 1927; (1925) — 164, 165, '184

— Adagio, vn, cl, pf [arr. Chamber Concerto: movt 2] (1956)

— Schliesse mir die Augen beide (Storm), 1v, pf, 2nd setting, 1925; pubd in *Die Musik*, xxii (1930); repr. (1955) — 165

— Lyric Suite, str qt, 1925–6; perf. Kolisch Quartet, Vienna, 8 Jan 1927; (1927) — 166ff, 182, 184, 190

— Three Pieces from the Lyric Suite, str orch [arr. movts 2–4] cond. Horenstein, Berlin, 31 Jan 1929; (1928)

— Der Wein (Baudelaire, trans. George), concert aria, S, orch, 1929; perf. R. Herlinger, cond. Scherchen, Frankfurt, 4 June 1930; vocal score (1930), full score (1966) — 150, 170, 171, 172

— Four-part Canon 'Alban Berg an das Frankfurter Opernhaus' (Berg), 1930 (1937)

— Lulu (opera, 3, Berg, after Wedekind: Erdgeist, Die Büchse der Pandora), 1929–35, orchestration of Act 3 inc.; cond. R. F. Denzler, Zurich, 2 June 1937; vocal score of Acts 1–2 (1936), full score of Acts 1–2 and excerpts from Act 3 contained in suite (1964); vocal score of Act 3 (1979); orchestration of Act 3 completed by F. Cerha; 1st perf., cond. Boulez, Paris, 24 Feb 1979 — 146, 147, 150, 165, 170, 171, 172, 173ff, 186, 187, 189, 191, 192

[5] Symphonische Stücke aus der Oper 'Lulu' (Lulu-Suite), S, orch; cond. Kleiber, Berlin, 30 Nov 1934; (1935) — 187, 189, 191

— Violin Concerto, 1935; perf. Krasner, cond. Scherchen, Barcelona, 19 April 1936; (1936) — 150, 171, 187, 188, 189, 191

Arrs. of F. Schreker: *Der ferne Klang*, vocal score (Vienna, 1911); *A. Schoenberg: Gurrelieder*, vocal score (Vienna, 1912), and *Litanei* and *Entrückung*, from Str Qt op.10, 1v, pf, 1912 (Vienna, 1921)

Principal publisher: Universal

WRITINGS

Arnold Schoenberg, Gurrelieder: Führer (Vienna, 1913)
Arnold Schoenberg, Pelleas und Melisande, op.5: thematische Analyse (Vienna, n.d.)
Arnold Schoenberg, Kammersymphonie, op.9: thematische Analyse (Vienna, n.d.)
Der Verein für musikalische Privataufführungen (Vienna, 1919) [prospectus]; Eng. trans. in Reich (1965)
'Die musikalische Impotenz der "neuen Ästhetik" Hans Pfitzners', Musikblätter des Anbruch, ii (1920); Eng. trans. in Reich (1965)
'Warum ist Schoenbergs Musik so schwer verständlich?', Musikblätter des Anbruch, vi (1924); Eng. trans. in Reich (1965)
'Offener Brief an Arnold Schoenberg' [on the Chamber Conc.], Pult und

Taktstock (1925), Feb; Eng. trans. in Reich (1965)
'A Word about "Wozzeck"', MM, v/1 (1927), 22; repr. in Reich (1965)
Lecture on Wozzeck [first given 1929], Eng. trans. in Redlich (1957)
'Die Stimme in der Oper', Gesang: Jahrbuch 1929 der UE (Vienna, 1929); repr. in Reich (1965)
Praktische Anweisungen zur Einstudierung des 'Wozzeck' (Vienna, 1930); repr. in Reich (1937); Eng. trans. in MT, cix (1968), 518
'Was ist Atonal?' [radio interview, Vienna, 23 April 1930]; Eng. trans. in N. Slonimsky, Music since 1900 (New York, 1938)
ed. H. Berg: Alban Berg: Briefe an seine Frau (Munich, 1965; Eng. trans., 1971)

194

BIBLIOGRAPHY

MONOGRAPHS AND COLLECTIONS OF ESSAYS

Musikblätter des Anbruch: Alban Bergs 'Wozzeck' und die Musikkritik (Vienna, 1926)

W. Reich, ed.: *23* (1936), nos.24–5 [special issue]

W. Reich: *Alban Berg: mit Bergs eigenen Schriften und Beiträgen von Theodor Wiesengrund-Adorno und Ernst Křenek* (Vienna, 1937)

H. Redlich: *Alban Berg: Versuch einer Würdigung* (Vienna, 1957; Eng. trans., abridged, as *Alban Berg: the Man and his Music*, 1957)

W. Reich: *Alban Berg: Bildnis im Wort* (Zurich, 1959)

——: *Alban Berg* (Zurich, 1963; Eng. trans., 1965)

T. Adorno: *Alban Berg* (Vienna, 1968, rev. 2/1978)

G. Ploebsch: *Alban Bergs 'Wozzeck'* (Strasbourg, 1968)

International Alban Berg Society Newsletter (1968–)

K. Schweizer: *Die Sonatensatzform im Schaffen Alban Bergs* (Stuttgart, 1970)

M. Reiter: *Die Zwölftontechnik in Alban Bergs Oper Lulu* (Regensburg, 1973)

M. Carner: *Alban Berg* (London, 1975)

E. Hilmar: *Wozzeck von Alban Berg: Entstehung–erste Erfolge–Repressionen (1914–1935)* (Vienna, 1975)

V. Scherliess: *Alban Berg* (Reinbek bei Hamburg, 1975)

E. A. Berg, ed.: *Alban Berg: Leben und Werk in Daten und Bildern* (Frankfurt, 1976)

P. Petazzi: *Alban Berg: La vita, l'opera, i testi musicali* (Milan, 1977)

K. Vogelsand: *Dokumentation zur Oper 'Wozzeck' von Alban Berg: die Jahre des Durchbruchs 1925–32* (Laaber, 1977)

E. Barilier: *Alban Berg: Essai d'interprétation* (Lausanne, 1978)

R. Hilmar: *Alban Berg: Leben und Wirken in Wien bis zu seinen ersten Erfolgen als Komponist* (Vienna, 1978)

O. Kolleritsch, ed.: *50 Jahre Wozzeck von Alban Berg: Vorgeschichte und Auswirkungen in der Opernästhetik,* Studien zur Wertungsforschung, x (Graz, 1978)

F. Cerha: *Arbeitsbericht zur Herstellung des 3.Akts der Oper Lulu von Alban Berg* (Vienna, 1979)

D. Jarman: *The Music of Alban Berg* (London, 1979)

G. Perle: *The Operas of Alban Berg*, i: *Wozzeck* (Berkeley, 1980)

R. Klein, ed.: *Alban Berg Symposion: Vienna 1980* (Vienna, 1981)

OTHER LITERATURE

E. Viebig: 'Alban Bergs "Wozzeck": ein Beitrag zum Opernproblem', *Die Musik*, xv (1923), 506

W. Reich: 'Aus unbekannten Briefen von Alban Berg an Anton Webern', *SMz*, xciii (1953), 49

K. Blaukopf: 'Autobiographische Elemente in Alban Bergs "Wozzeck" ', *ÖMz*, ix (1954), 155

J. Russell: *Erich Kleiber* (London, 1957)

G. Perle: 'The Music of *Lulu*: A New Analysis', *JAMS*, xii (1959), 185

——: *Serial Composition and Atonality* (Berkeley, 1962, rev. 5/1981)

R. Offergeld: 'Some Questions about *Lulu*', *HiFi/Stereo Review*, xiii/4 (1964), 58

G. Perle: '*Lulu*: the Formal Design', *JAMS*, xvii (1964), 179

I. Vojtech: 'Arnold Schoenberg, Anton Webern, Alban Berg: unbekannte Briefe an Erwin Schulhoff', *MMC*, xviii (1965), 30–83

M. DeVoto: 'Some Notes on the Unknown Altenberg Lieder', *PNM*, v/1 (1966), 37–74

G. Perle: 'Erwiderung auf Willi Reichs Aufsatz "Drei Notizblätter zu Alban Bergs *Lulu*', *SMz*, cvii (1967), 163

——: 'Die Personen in Bergs *Lulu*', *AMw*, xxiv (1967), 283

——: 'Die Reihe als Symbol in Bergs *Lulu*', *ÖMz*, xxii (1967), 589

B. Archibald: 'The Harmony of Berg's "Reigen" ', *PNM*, vi/2 (1968), 73

G. Kassowitz: 'Lehrzeit bei Alban Berg', *ÖMz*, xxiii (1968), 323

D. Jarman: 'Dr. Schön's Five-Strophe Aria: some Notes on Tonality and Pitch Association in Berg's *Lulu*', *PNM*, viii/2 (1970), 23

——: 'Some Rhythmic and Metric Techniques in Alban Berg's *Lulu*', *MQ*, lvi (1970), 349

N. Chadwick: 'Berg's Unpublished Songs in the Österreichische Nationalbibliothek', *ML*, lii (1971), 123

——: 'Franz Schreker's Orchestral Style and its Influence on Alban Berg', *MR*, xxxv (1974), 29

H. Knaus: 'Studien zu Alban Bergs Violinkonzert', *De ratione in musica: Festschrift Erich Schenk* (Kassel, 1975), 255

——: 'Berg's Carinthian folk tune', *MT*, cvii (1976), 487

L. Treitler: ' "Wozzeck" et l'Apocalypse', *SMz*, cvi (1976), 249 [Eng. version in *Critical Inquiry*, winter 1976]

D. Green: 'Berg's *De profundis*: the Finale of the Lyric Suite', *International Alban Berg Society Newsletter*, no.5 (1977)

——: 'The Allegro misterioso of Berg's *Lyric Suite*: Iso- and Retrorhythms', *JAMS*, xxx (1977), 507

S. Jareš: 'Inscenace Bergova Vojcka v Národním divadle roku 1926' [The staging of *Wozzeck* in the Prague National Theatre 1926], *HV*, xiv (1977), 271

V. Lébl: 'Případ Vojcek' [The fall of Wozzeck], *HV*, xiv (1977), 195–229 [with Ger. summary]

Bibliography

G. Perle: 'Berg's Master Array of the Interval Cycles', *MQ*, lxiii (1977), 1–30

——: 'The Secret Programme of the Lyric Suite', *MT*, cxviii (1977), 629, 709, 809 [also in *International Alban Berg Society Newsletter*, no.5 (1977)]

——: *Twelve-Tone Tonality* (Berkeley, 1977)

——: 'Inhaltliche und formale Strukturen in Alban Bergs Oper "Lulu"', *ÖMz*, xxxii (1977), 427

F. Herschkowitz: 'Some Thoughts on *Lulu*', *International Alban Berg Society Newsletter*, no.7 (1978), 11

D. Jarman: '*Lulu*: the Sketches', *International Alban Berg Society Newsletter*, no.6 (1978), 4

J. A. Smith: 'Some Sources for Berg's "Schliesse mir die Augen beide" II', *International Alban Berg Society Newsletter*, no.6 (1978), 9

G. Perle: 'The Complete *Lulu*', *MT*, cxx (1979), 115

——: 'The Cerha Edition', *International Alban Berg Society Newsletter*, no.8 (1979), 5

D. Harris: 'The Berg–Schoenberg Correspondence: A Preliminary Report', *International Alban Berg Society Newsletter*, no.9 (1980), 11

G. Perle: ' "Mein geliebtes Almschi ..." Briefe von Alban und Helene Berg an Alma Mahler Werfel', *ÖMz*, xxxv (1980), 2

——: 'The Film Interlude of *Lulu*', *International Alban Berg Society Newsletter*, no.11 (1982), 3

D. Jarman: 'Alban Berg, Wilhelm Fliess and the Secret Programme of the Violin Concerto', *International Alban Berg Society Newsletter*, no.12 (1983), 4; *MT*, cxxiv (1983), 218

197

Index

Aber, Adolf, 163
Adler, Guido, 6, 89
Adler, Oscar, 2
Adorno, Theodor W., 147, 167
Altenberg, Peter, 138
Amsterdam, 12, 156
Archibald, Bruce, 112

Babbitt, Milton, 123
Bach, David Josef, 2
Bach, Johann Sebastian, 50, 59,
   68, 116, 117, 119, 188
Baden-Baden, 169
Bad Ischl, 90
Balzac, Honoré de, 45, 46
Barcelona, 14, 25
Bartók, Béla, 169
Baudelaire, Charles, 170
Bayreuth, 89
Beethoven, Ludwig van, 50, 58
Berg, Alban, 6, 7, 67, 68, 90, 91,
   100, 106, 122, 124, 137-97
Berg (née Nahowski), Helene
   [Berg's wife], 143, 152, 154, 175,
   192
Berg, Smaragda [Berg's sister],
   175
Berghof, near Villach, 137, 155
Berlin, 5, 8, 13, 14, 15, 25, 89, 90,
   143, 144, 163, 164, 169
——, Prussian Academy of Arts,
   13, 186
——, Staatsoper, 163
——, Stern Conservatory, 5, 8
——, Überbrettl, 5
Bethge, 104
Boston, Malkin Conservatory, 15
Boulez, Pierre, 123
Brahms, Johannes, 2, 29, 35, 59,
   91, 93

Bruckner, Anton, 19, 58
Bruck an der Leitha, 10
Büchner, Georg, 151, 152, 154,
   156, 161, 173
Busoni, Ferruccio, 13
Byron, Lord George Gordon, 62

Cage, John, 123
Cerha, Friedrich, 192
Chicago, University of, 17
Cologne, 92

Dallapiccola, Luigi, 123
Danzig, 90
Debussy, Claude, 154
Dehmel, Richard, 5, 30, 45, 94, 95
Diez, Ernst, 89

Eisler, Hanns, 13

Feldman, Morton, 123
Fliess, Wilhelm, 190
Forte, Allen, 119
Frankfurt, 92, 163
Franzos, Karl Emil, 156, 157
Freud, Sigmund, 190
Fröhliches Quintet, 4
Furtwängler, Wilhelm, 14, 67
Fuchs-Robettin, Hanna, 167, 190
Fux, Johann Joseph, 20

George, Stefan, 39, 40, 95, 123
Gerhard, Roberto, 13, 14
Gerstl, Richard, 37
Goethe, Johann Wolfgang von,
   107, 125
Graedener, 89
Graz, 89
Greissle, Felix, 5
Gropius, Manon, 187

198

Index

Index

Wood, Sir Henry, 9
Worther, Lake, 137

Zemlinsky, Alexander von, 2, 3, 6, 7, 9, 19, 31, 32, 47

Zemlinsky, Mathilde von: *see* Schoenberg, Mathilde
Zillig, Winfried, 13
Zurich, 191
Zweig, Stefan, 137, 138, 175